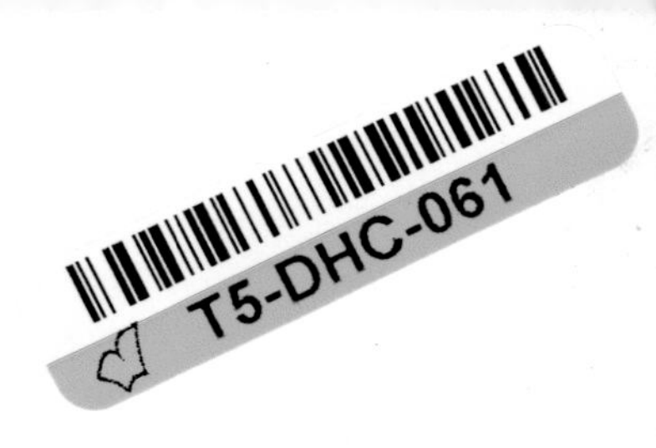

Dedicated to the pioneers
who came in covered wagons.

CONTENTS

CHAPTER 1: Indians

Stolen Sisters

This story has been handed down in my husband's family for years. It is about the miraculous escape from death at the hands of the Indians of the little girl who became my husband's great-grandmother.

The pioneer family lived in a sparsely settled country where roving bands of Indians often stole stock and food. Supplies were low and the father knew he must leave his family and make the long trip to the nearest town for groceries.

Even tho he pushed the team hard all day, it was a long day's journey. The heavily-loaded wagon was even slower on the return trip. It was very late the second night when he returned.

To his horror, he saw that where his cabin had stood now lay only a pile of glowing embers. He ran here and there frantically calling for his wife and children. One by one, he discovered the bodies of his wife and boys, cruelly murdered with their scalplocks torn away. The Indians had found his home unguarded, had burned and murdered and had driven his horses and cattle away.

A long, tedious search revealed no sign of his two little girls. So, as dawn broke, revealing the trail the Indians had taken, a broken-hearted father started out to try to find some trace of his two missing children.

One of the little girls had fought and cried until she became such a nuisance that the Indians killed and scalped her before her sister's frightened gaze. They threw her body beside the trail where her father later found it.

The other little girl wisely went along, putting up no resistance and biding her time. They traveled all that day and part of the night and then stopped for a short rest. The next day they pushed on, and the Indians became wary about covering their trail. After the second day's travel, they were very tired and stopped to sleep. One Indian was set to watch, and the others soon were fast asleep.

The little girl watched closely, tho she pretended to be asleep. When the guard's head fell forward on his chest, she slipped quietly out of camp. Every step she took, she expected to feel a cruel hand press her shoulder. But when she was safely out of camp, she ran wildly, frantically, paying no attention to direction. Her only thought was to put as many steps between herself and her captors as possible. Finally, exhausted, she fell to the ground, gasping for breath.

Just as we teach our children what to do in case of fire, those pioneer children had been taught to go eastward if they were lost. Westward was wilderness; but to go eastward meant that one would eventually find a white settlement. As she collected her wits, this is what the little girl did. She traveled mostly at night. She found a few patches of wild berries; she drank from streams. Weary and footsore, she stumbled into a white settlement at last.

Her father had lost the trail and finally had turned sorrowfully back to bury the other members of his family. Means of communication were very slow in those days and it was quite a long time before he was reunited with his little daughter.

Mrs. F. J. Testerman
Vera, Okla.

Settlers in a Fort

Indian scares plagued the early settlers in southern Kansas. Some men paid Chief Chetopa five dollars a year, and he and his tribe never molested them. However, rumors of attack persisted. At one time the settlers in my own locality deemed it well to gather in a home built of logs like a fort.

Five families crowded into two small rooms, children crying, women half-hysterical, and the men with guns in their hands. Late in the night the hostess suggested that they make coffee for the men. "But," said she, "we'll have to shell and parch the corn because we have no real store coffee."

"Let's not bother to shell it," said a neighbor woman. "Just throw it in the b'iler, cob and all—we'll all be scalped before morning anyhow."

But the corn was shelled, parched and brewed. It tasted good, and morning dawned peacefully. The settlers returned to their homes and never saw a really hostile Indian then or thereafter.

Mrs. Pruda B. Utley
Arkansas City, Kan.

Scalping of Eddie Malone

When my grandparents were pioneers near Salem, Neb., there was a dependable boy named Eddie Malone in the neighborhood. When some surveyors came thru headed west they hired Eddie to ride ahead of them on a mule and "set stakes." One day Eddie had gone ahead over a hill in what is now western Kansas. When the surveyors got to the top of the hill, they froze in horror.

A band of Indians in war paint had surrounded Eddie. The frightened boy urged the mule to run and break thru the circle, but when the mule tried it, one of the Indians hit it across the nose and grabbed Eddie. They dragged him from the mule and scalped him. Then they placed the scalp on an arrow and waved it at the surveyors to warn them. They could do nothing but leave. Eddie was dead, and they would be dead, too, if they tried to go ahead with their survey.

Mrs. Flora Rorabaugh
Box 122
Norcatur, Kan.

Little Belle

My grandparents started from Kansas to Oregon in a covered wagon in 1876 with their seven children. Little Belle, a two-year-old with golden curls, was their pride and joy. They followed the Oregon Trail over mountains and bridgeless rivers.

In Colorado they stopped for a few days to rest the horses and do the washing. One day some Indians stopped to stare at them. They never had seen a white child with golden hair before. They wondered if baby Belle was real and they wanted to touch her. They offered to buy her, trade a horse for her—and at last they offered a quarter of dried dog meat for her! They were angry when Grandmother turned down all their offers.

The next day, little Belle went out to play and wandered away from camp while everyone was busy. When it was discovered that she was gone, the whole family hunted for her. They feared the Indians had stolen her. After much grief and worry, she was found asleep under a sage bush.

How Grandmother praised and thanked God for protecting her baby!

Aunt Belle is still alive and is 81 years old and quite feeble, but she has many memories of pioneer days in Oregon.

Mrs. J. S. Thompson
Auburn, Wash.

Hunting Knife Haircut (Reprinted from Capper's Weekly)

We girls were discussing hair styles when my husband's grandmother spoke up and said, "I only had my hair cut once in my life."

We all wanted to know when because we couldn't imagine Grandma with short hair. She told us it was back when she was a little girl of six or seven. The settlers had been warned that an Indian party was on the way.

"Our papa told us," said Grandma, "that if we girls had our hair short and ugly the Indians wouldn't be as likely to scalp us as they would if it was long and pretty. Then he took his hunting knife and whacked off our long locks."

Grandma went on to explain that the alarm was a false one, but that she'd never forgotten the one time she'd had short hair. Guess in those days a man's hunting knife had lots of uses.

Mrs. Jack Hurd
Bolivar, Mo.

Regina, Mother of Pioneers

In 1755, so the story handed down in our family goes (and history books and a monument erected in Pittsburgh, Pa., to honor her confirm it), my maternal grandmother seven generations back was carried off by the Indians after they had killed her father and brothers and burned her home.

Great-great-great-great grandmother's name was Regina Hartman; the home the Indians ravaged was near the eastern boundary of Pennsylvania. Regina and her sister were hurried off and made to travel fast day and night thru woods and over rocks, up and down hills. A third and smaller girl the Indians had taken captive clung to Regina as to a mother. Altho Regina was only nine, she had to carry the weary child a good part of the way. When they finally came to a halt the sister was taken on, never to be heard from again. Regina worked hard for the Indians during the nine years she was held captive and became as one of them.

After the battle of Bushy Run, her mother sought the girl when the released prisoners were taken back to Carlisle, Pa. She found a swarthy maiden who could talk only in the Indian tongue. Efforts at recognition failed. Then the mother thought to sing a song she had taught Regina as a tiny tot. The girl, with memories of her childhood revived, took up the refrain, sang it thru, then repeated Luther's Catechism as she had learned it in her home, and rushed into her mother's arms.

Thirty years later Regina's daughter Maria and her husband, John Adam Baker, loaded their three children and a few choice possessions into a covered wagon. Leaving their Northampton

County home, in eastern Pennsylvania, they started on the long and weary journey westward across the state, wending their way over mountains where bears and other wild animals were often their only company. The trail they took doubtless followed the one of the forced march the Indians had taken Regina over years before, so they were ever on the alert for treacherous red skins. This was about 1795, when travel for many between Philadelphia and Pittsburgh was on foot, which was the means of delivering messages. The State road had been built at a cost of $20 a mile, so it is likely they did not do much more than cut down the trees. Isn't it interesting to compare that price with the present rate per mile of road building?

Despite the hazards of travel, the little family made the trip safely, came at last to Westmoreland County, where they bought land south of Greensburg, about 40 miles from Pittsburgh. There they built a log cabin, settled down to grow up with the country, propagated and prospered for at least 80 or 90 years before the wanderlust again set in.

In the mid-seventies "Kansas Fever" struck western Pennsylvania like the plague. Inoculation was unknown those days, so the westward trek was on! Many came by covered wagon. Not so my father, Edgar Clark Fowler.

As a young man of 20, in the fall of 1877, he packed his traveling bag. Grandmother Fowler filled another suitcase full of food for him, and he bought a one-way railroad ticket, via St. Louis and Kansas City, for Ottawa, Kan. There were a few Indians in Dad's story too, but they were mostly in the minds of fond relatives and friends who gathered at the little railroad station to see him off. With tears in their eyes they bade him a last farewell, so sure they were that he'd be tomahawked way out there in Indian-ridden Kansas.

Trains were slow those days, but there was still food in the big valise when the engine finally chugged into Ottawa. There, young Ed Fowler built himself a wagon, bought a team of horses, and drove the Kaw River road thru Lecompton to Topeka. He put up temporarily at the old Gordon House, the town's deluxe hostelry

They were only a few miles from Ft. Wallace and at the present site of Scott City, Kansas. They had one wagon, six oxen and one horse. They had left a wagon train of home seekers a few days before. They were from the Blue Ridge Mountains in the South and instead of taking up a Kansas homestead, they were headed for Colorado where they heard there were streams and trees.

It was a tragic error in judgment. They left their campfire burning that night, and the next morning Indians swooped down on the little camp. No tragedy of the plains ever exceeded in horror the raid of the Cheyenne Indians on the German family.

This family was a sturdy one, fitted for pioneer life in the West. John German and his buxom wife had one son, Stephen, 21, and six daughters, Rebecca, 23; Catherine, 19; Joanna, 17; Sophia, 15; Julia, 10; and Adelaide, 5.

Three of the girls were mature young women, all pretty and well-formed. Catherine was considered the prettiest of the six. Sophia, too, was a winsome girl, and Joanna had long, luxuriant, curly hair that fell in shimmering beauty about her shoulders.

It was the custom of wagon travelers to rise early and get started by daybreak. The Germans were ready to be on their way by sunup. Stephen was out a little way from the wagon to try to get a prairie chicken or a rabbit for their supper that night when the family heard the shrill, blood-thirsty yell of Indians.

Up from the gulch behind them rode the red warriors. Young Stephen was shot down before he could fire one shot. Vigorous John German fell next with a bullet thru his head.

Forgetting that the savages could not understand her language, Mrs. German ran to her husband and began to plead for mercy. A brawny Indian grabbed her arm, swung her around and plunged a butcher knife in her side. She gasped and died.

When her father was shot down, Rebecca Jane jumped from the wagon and with Amazon-like strength attacked the Indians. She sank her ax blade in a warrior's shoulder, but was shot in the back by another savage.

In an incredibly short time only five of the family were left. The savages stopped their slaughter for a time and stood regarding the girls. Either the Indians did not want more than four captives, or else Joanna's beautiful long hair, which would make a showy scalp, decided her fate. For a time she was left sitting on a box surrounded by savages while the four other girls were carried away in the arms of the Indians.

As the tribesmen galloped away with the captured girls, a shot was fired. Her sisters did not know it ended the life of Joanna until a savage loped up and, with a hideous grin, displayed a bedraggled scalp that was recognized because of the lovely long hair.

Mile after mile the terrified girls traveled in the clutches of the warriors, who took them to the Indian village of Chiefs Greybeard and Stone Calf. The warriors were rewarded by being allowed to take the older girls to another part of the camp where they suffered the fate of most white women who fell prey to Indians.

Shortly afterward, Stone Calf took his band and the older girls, Catherine and Sophia, and moved to Texas. Greybeard kept the little girls, Julia and Adelaide, and gave them to a niece who had no children. Altho the squaw had asked for the children, they suffered greatly.

When Gen. Nelson A. Miles heard of the massacre of the German family, he took steps at once to recapture the girls. When it became known that Greybeard's village was where the small girls were held, he sent Greybeard a warning. The chief then deserted the children on the prairie because he was afraid of being caught with them.

For almost three months the little girls wandered over the prairie eating wild grapes and hackberries found along the streams and sleeping in the tall grass and brush. The country was full of wild game at that time or they probably would have been eaten by the wolves.

One day a roving band of Indians found the children and took pity on them and took them again to Greybeard's camp. Soon afterward, Lieutenant Baldwin and a detachment of troops made a surprise attack on the village and rescued the children who were gaunt, half-starved and bruised. Dr. James L. Powers, the medical strange reason, Sister was very much afraid of feathers as they moved about in the air. Mother discovered she had only to put a large feather in the unscreened open doorway and the baby would not venture into the yard.

Ella Bedsaul
La Mesa, Calif.

Mountain Road Adventure

When we went from South Texas to Colorado in covered wagons, we sometimes would travel for days without seeing a house. When we got into the mountains the roads were so steep and narrow that Father would have us all get out of the wagons and walk until the road got wide enough for us to ride again without his being afraid the wagons would upset.

One day when he had us walk we got away behind the wagons. We were loitering along, carefree and happy, when we looked up the side of the mountain. There, so plain I can see it to this day, was a big mountain lion lying in the mouth of a cave!

We ran in pure panic, and never have you seen such breathless, frightened children as we were when we caught up with the wagons.

Mrs. Lee Johnson
Mangum, Okla.

Feared Coyotes

My mother-in-law came to the wheat land in a covered wagon as a wife of 17, cradling a six-week-old baby in her arms. No one will ever know the hardship and suffering she endured. She fought prairie fires with the men, worked in the fields, kept the home and lived long enough to give birth to six more sons, alone and unattended.

One day her two little boys, ages two and three, wandered away into the tall wheat and became lost. Tho the parents searched and called, no trace of them could be found. By evening it had started to rain and darkness settled down. The men gave up the search until morning.

"But I couldn't sleep," Ma said. "I stayed up all night with the light in the window and prayed that the coyotes wouldn't kill them."

The next morning the two little brothers came across the yard, soaked to their knees from the wet grass and too little to tell what had happened. During the wheat harvest, the workers came across an old wagon bed upturned in the corner of a field about a mile and a half from the house. There was a little rounded out nesting place inside. The men thought they had solved the mystery of little lost boys. Mother only knew that her prayers had been answered that summer night.

Mrs. C. O. Barnes
Redlands, Calif.

Panther and a Moon-eyed Horse

Grandma used to tell a tale that always held us children spellbound.

When she was first married, she had an old moon-eyed horse. She rode the horse one evening to the home of a neighbor to get a setting of eggs. It was a distance of two miles, and half of it was thru a dense woods.

It wasn't quite dark when she went over, but by the time she started back it was cloudy and very black. When she entered the pitch-dark woods on her return trip, she heard the scream of a panther close behind her.

The horse was as frightened as Grandma, and he stumbled thru the brush and under tree limbs with Grandma hanging on for dear life. The horse could not see well and kept getting off the trail. The panther could be heard getting closer and closer as he crashed thru the brush or jumped from tree limb to tree limb.

When Grandma and the horse finally reached a clearer area, the panther stayed behind. Grandma's long hair had fallen down and was snarled with twigs and brush. Her hands and face were badly scratched. And the eggs? They were still in the basket with the handle looped over her arm, but there wasn't a good one left. I guess Grandma was lucky the horse couldn't see well, or she never would have been able to stay on.

Miss Velma Sipes
Unionville, Mo.

Animal Fury

I am 81 years old and I can remember many things from the ox team and covered wagon days. One of the ideas people have now is that things were pretty slow and dull then, but I know we had excitement every time we drove an ox team to church and had to pass cattle on the way. The herds of cattle would come on a run to the fence to see the strange oxen on the road and would cause us trouble.

One time we were going to a picnic and had to pass a herd of cattle. The bull jumped the fence and came pawing and bellowing to fight our oxen. My father unhitched the chain but left the oxen yoked because they could fight better in the yoke. He had a heavy ox whip and he stood by with it to help his team.

There was half a wagonload of us children. We jumped down and ran around like scared chickens. I have no idea how it would have ended if the bull's owner had not come riding up at a hard gallop with his big bull whip. My, were we relieved! Yes, we had excitement even with a dull, plodding team of oxen.

Mrs. T. K. Mannon
Brown Branch, Mo.

Bears and Bobcats

Bears and wolves killed livestock and rubbed and scratched around the log cabins in the heavily wooded area of Wisconsin where both my father and my mother came with their parents in covered wagons. Bobcats and panthers were numerous and would spring from the trees onto cattle.

My parents grew up in this wild country and were married there. Later they lived a few years in Iowa. In 1878 they heard about some new land being opened to homesteaders in South Dakota. They started for this new frontier in two covered wagons loaded with their eight children and all their possessions, including a spinning wheel.

A prairie fire had gone thru the country and ashes, sand and high winds made travel difficult. They hurriedly broke sod and laid up a house and a shed for stock. Two days after the roof was on, there was a terrible blizzard. Drifts were 12 feet high and many lives were lost. My oldest brother froze his feet and it was thought for a time he would lose them. Many lives were lost from lack of medical care. The nearest doctor was 50 miles away.

Three of the things people feared most in the raw, new country were uncivilized Indians, horse thieves and wild animals!

Mrs. Emma Michaelsen
Mesa, Colo.

had submitted to. Across the creek he thought he saw a dark object. Excitedly, he shook Huldah back to unwilling consciousness and they got woodenly to their feet. Praying for strength, they nerved themselves into what seemed a final effort to live. They started for the creek, the snow crunching angrily under their feet. Testing each stiff step, they crossed the frozen surface of the stream and sought the dark object. It was a haystack with cattle nearby.

Dawn came quickly and revealed a little path that led to the dugout they had set out to find 14 hours before. The family was up and had a fire. They answered David's rap and were surprised to receive a New Year's call from their pastor and his wife so early in the morning.

Mrs. Louis Grimm
Wauneta, Neb.

Mules Shared Shelter

My father, his brother and two neighbors left Missouri to take government claims near what is now Ness City, Kansas. They staked out adjoining claims and all worked together to get one sod house built before winter caught them.

They had just completed the house when a big blizzard swooped down. They brought all eight head of mules in with them to keep them from freezing. They took off the wagon box and put it between mules and men to keep from being kicked to death. They ate and slept that way for three days until the storm blew itself out. They had to scoop snow as high as the door to get out.

Mrs. Elda Whitney
Sheridan, Mo.

Survived on Thistles

My father was a well driller in Kansas in the 1890s and often was away from home, leaving us to care for the stock. Once when he was gone, there came a terrible blizzard. My mother didn't dare venture out of the house to see about the stock.

After three days, the storm ceased and we could see that the pole stable was drifted completely over. Mother walked on top of the drift and tore a hole in the shed roof until she could see that the stock was safe. Father had cut green Russian thistles and stacked them against the stable for a windbreak. The cattle had torn down their mangers and lived on that thistle hay!

After that, whenever we had a dry year, Father cut and stacked the thistles and they tided the cows over so the children had milk, even tho other food was scarce.

Mrs. James Eggers
Box 1321
Cortez, Colo.

Burned Corn for Fuel

Here is my blizzard story as my mother told it to me:

"My parents were living in western Kansas in a large one-room cabin with walls made of rough plank and no ceiling overhead. One day my father got up before daylight and drove across the plains to buy a load of corn for his stock. He didn't get back until after dark, but the sky was clear and the stars were shining.

"About midnight the family was awakened by a howl of coyotes and the roaring of the wind. They had no wood to burn and the cold was so bitter they knew they might be facing death. They had a big canvas tent and they stretched it across the room to hold out some cold, but still they suffered.

"The cold wind raged on and on and they started burning the golden ears of Kansas corn. It was two days and nights before the snow stopped. By that time they had burned all of the corn, but they had saved their lives and those of their two babies.

"On the third day a neighbor ventured out to search for his stock. He came to my mother's place nearly frozen. They took him in and fed him and warmed him up. When he started home, he asked my mother to let him borrow her old-fashioned bonnet to protect his head and face from the wind and she let him have it gladly."

I think my mother mentioned this every time we had a cold winter for as long as she lived.

Mrs. Charles R. McKnight
Chamois, Mo.

CHAPTER 4: Prairie Fires and Plagues

Wall of Flame

The little homestead shack was by itself on the lonely prairie, and on this warm, sultry day it seemed more isolated than usual among the acres of grass. Mama and Papa had gone the 14 miles to town and we girls were home all alone.

The two older girls were doing some baking and cleaning and I was outside playing when all at once I noticed a huge black cloud in the west. My cry of "Prairie fire!" brought the girls on the run. By this time the flames were visible and the fire line appeared to be a mile or more long and headed straight for our home.

Quickly my sisters decided it was too late to try to throw up any protection. Anyway, the team was gone and only Old Maud was in the corral. My older sister caught Maud and with clumsy efforts managed to get the single harness on her. The three of us got the buggy shafts and tugs where they belonged. Into the buggy we piled along with a few valuables the girls had grabbed in the house.

By this time the fire was less than a mile away and coming rapidly toward us. The girls decided to drive away from the flames. When we had driven a short way we came to a ditch, dug in the hope of future irrigation. There was no bridge, but Old Maud scrambled across—buggy, girls and all. Only then did we stop and take time to look back.

We could see movement in the smoke just in front of the flames. What was it? Presently we could see smoke-blackened fire fighters, men from a nearby government camp and homestead menfolks. They

on our claim in Butler County, Kansas. In the spring of 1874 my father had put in corn and vegetables.

The corn had done so well and was in roasting ears when one day a black cloud came from the south so that we could not see the sun. It was grasshoppers and they covered the ground four inches deep in a very short time. Nothing was left but the cornstalks. They ate shucks and there was not a leaf or a blade of grass.

Leighton W. Flock
Nampa, Idaho

Hungry Hordes of '74

My father landed in Kansas a day's drive west of Salina on July 1, 1874. He bought 40 acres of land with a dugout on it and a small field of corn in the roasting-ear stage. On July 16 the grasshoppers came in a black cloud from the northwest. By morning every green thing was gone. My father picked up buffalo bones and hauled them to Salina and sold them to buy feed for the team and groceries for the family. I am 87 years old.

Nora Tinsley
El Dorado Springs, Mo.

Ravaged the Earth (Reprinted from Capper's Weekly)

I remember Mrs. H, a pioneer woman who weighed less than 100 pounds, but whose ambition and courage were unmatched.

She brought two lilac bushes from her fine home in Virginia to the barren plains of Nebraska when she came in a covered wagon. She often went without a cool drink herself so her lilacs could have water, and the bushes thrived in spite of hot winds and cold winters.

Then came the terrible grasshopper hordes. They could be heard long before they arrived. Mrs. H tied up one bush in her beautiful bedspread, the only one she had. The other bush was covered with a huge buffalo robe.

The grasshoppers were so thick they blackened the skies like a storm, and they ate their way thru everything. Hours later, Mrs. H went out to find everything gone—buffalo robe, bedspread and bushes. Only bare stumps were left in the ravaged earth.

The next spring, however, the lilacs came up from the roots, more beautiful than ever.

Mrs. Fred Walter
Wallace, Neb.

CHAPTER 5: Doctors Were Far Away

Alone in a Strange Land

It was arranged that if Mother needed Father when he was working in the field, she was to tie a red flag on a pole. One day on our pioneer homestead (where our taxes the first year were 47 cents) one of the four small children became very ill suddenly.

When Father saw the red flag, he rushed to the house and started on horseback for the doctor. He was gone nearly all day. When Father and the doctor got back, the baby had been dead for six hours.

My Mother felt so alone in a strange land that day as she sat by the body of her baby. Such were the ways our pioneer trails were blazed. Thanks to the tenacity of our parents, we are here to tell the stories.

Mrs. Henry M. Price
Lawrence, Kan.

Birth in a Covered Wagon

This story was told to me by Aunt Matilda:

"My husband was very impatient to get started for the West where lots of free land was to be had. In 1882 we started from Illinois to Kansas. We had two children, eight and six years old, and I was expecting another arrival. My husband told me we could make it in time for our baby to be born in our new home.

"The days were long and the journey rough as we picked our way in the jolting wagon. Sometimes there was a faint trail. More often, there was nothing. As we neared eastern Kansas, I knew my time had come.

"There was no doctor or midwife to help me, and with only the help of my husband, my baby girl was born in the covered wagon. No one will ever know the agony I suffered. I prayed to God to let me die. But, somehow, I pulled thru and have lived to a ripe old age."

Mrs. J. Singley
Marion, Texas

Death in a Covered Wagon

My grandmother was born in Kentucky in 1815. When she was four, her parents decided to move by covered wagon to the new land of Missouri. My grandmother's grandmother left all that was familiar to her to go with her daughter's family.

As they followed the trail thru southern Illinois, the old lady died. The wagon train stopped and the men of the party cut a huge tree. As carefully as possible, they cut a slab off one side and hollowed the tree trunk as for a dugout canoe. Then they laid my great-great-grandmother's body in the shell, carefully laid the slab back and nailed it down. They dug a grave in the trail, lowered the improvised casket, read the Bible and prayed. They flattened the grave after they had filled it and drove back and forth over it with the wagons so that any wandering Indians might not be able to tell that anyone was buried there.

My grandmother's one vivid memory of the whole trip was of her mother standing at the back of the wagon staring back toward the spot where they had left her mother's body for many hours after the spot was out of sight.

Margaret Harris Heck
Quapaw, Okla.

Baby's Snake Bite

In the 1870s my grandparents lived on a homestead. One evening their little two-year-old boy was bitten by a rattlesnake as he reached under a pile of wood to pick up his kitten.

Grandfather was miles away breaking prairie with oxen, and Grandmother knew he would not be home that night. She hurried to the spring and got a handful of ice-cold mud. On the mud she poured some vinegar and soda. She applied this to the snake bite on the baby's hand.

The child grew very, very sick, but all night long Grandmother kept up the treatments. Early the next morning she started out to go thru the woods to the home of relatives five miles away. There were no roads and she stumbled thru the woods carrying the sick child. It was June and the sun was very hot. The relatives tenderly cared for the baby and for her. He recovered and lived to be an old man. Grandmother lived to be an old woman and often told this story to her grandchildren.

Mrs. Giles Cleveland
Lyons, Neb.

700 Miles With a Sick Baby

The popular slogan around 1875 was "Go West and make your fortune." My father was intrigued with the idea, and he traded our farm on the banks of the Wabash in Indiana for a farm in Kansas, ten miles from Independence, which at that time was an Indian trading post. Our farm was located on the trail from the Indian Territory to Independence.

Father sold all our livestock, farming implements and furniture and we were going by train. However, when we went to spend the last night with our maternal grandmother, my three-year-old brother took meningitis and lingered for weeks between life and death. When he began to improve, the doctor advised that the trip be made by wagon, so Father bought a wagon and a span of mules.

We started on the 700-mile trip on October 6, 1878, on my fifth birthday. Altho I'm in my 82nd year, many of the incidents stand out clearly in my mind. We were on the road for seven weeks.

When we crossed the Mississippi at St. Louis, our team was frightened by a train and ran away. One of the rear wagon tires came off and, as you oldsters know, the tire had to be heated before it could be put back on the wheel. Father started a fire along the road. Soon a number of men came from a nearby saloon and offered their assistance. Father accepted gladly because it was no job for one man.

Imagine the feelings of my mother when she left her comfortable home on the Wabash and had to care for a sick child traveling 700 miles by wagon. When we reached our destination, we found a one-room log cabin with a rock floor instead of the good four-room framed house the agent had described!

However, I do not remember that my mother ever rebuked my father for the arduous trip and unprofitable move.

Mrs. Blanch Camp
Pico, Calif.

Saved Her Son's Thumb

One time Grandfather was gone and the two little boys were helping Grandmother chop wood. George chopped Fred's thumb, leaving it hanging by the skin. There wasn't a doctor for many, many miles.

Grandma poured sugar on Fred's thumb, put it back together and bandaged it. I don't know why she put sugar on it, but Uncle Fred kept his thumb, tho it was a bit stiff all his life.

Mrs. Robert Williams
Winfield, Kan.

Amputated Frozen Fingers

In November, 1872, when I was three months old, my parents came to Kansas in a covered wagon from Iowa. Father often left home to hunt buffalo. The best hunting ground was where Pratt now stands.

Several men went during a very severe winter and one man froze his fingers so bad his wife had to amputate the ends of two or three of them herself. There was no doctor in miles.

They would skin the buffalo and cut out the choice meat from the hind quarters to bring home. They took the hides to Hutchinson and sold them for $1. Father would bring home enough supplies to last and then he would be out on the range again.

My mother was 27 years old and had three children. I've heard her tell how my brother and sister would sit on a box (we had no chairs) with their feet in the oven. They would burn the toes of their shoes and their heels would still be cold! She said they could look out the cracks of the house on a moonlight night and see coyotes sitting in the snow.

Mrs. Z. Smith
Wakita, Okla.

Death from Exposure

Since I will never have any grandchildren to relate the story to, I'd like to tell you of an arduous journey in a covered wagon that I took when I was a little girl.

My parents with three children in one wagon and my married sister and her husband in another left South Dakota for Arkansas. As we were going thru Iowa, we were caught in a blizzard. The folks made camp, and Father and Mother sat up all night burning corn in our little cookstove to keep from freezing.

In Missouri we were almost drowned in a flash flood. We had camped over night by a creek. Some time during the night we were awakened by the noise of buckets and pans banging around. Before the men could get the horses hitched to the wagon, the water was up to the hubs. We drove to high ground and set up our tent on a hillside. It rained for days. My mother gave birth to a baby girl on the 23rd of December. As soon as the high water went down and Mother was able to travel, we resumed our journey. She carried the baby on a pillow. In January our baby sister died and our Father had a long sick spell—both because of exposure in the cold, wet weather. My married sister died a short time later from a cough brought on by the exposure on our covered wagon trip.

Rose Brown
Louisburg, Mo.

Lost Their Baby

The year was 1869. Mother was only nine years old. Grandfather and Grandmother loaded their four children and their few possessions into the wagon and left Kentucky, prayerfully headed westward.

Their journey across rough trails was a slow one. Winter set in early. One cold night God called their youngest child to Him. They knelt beseechingly to the Father for His help. Late that night they dug the little grave and tenderly laid their baby to rest. They prayed and sang "Rock of Ages." The rest of the night was spent driving the wagon and oxen back and forth over the grave. This left no trace for Indians who might dig up the body.

They reached Cottonwood Falls, Kan., where they stayed until spring. Then they staked their claim four miles east of Wichita and built a sod shanty. When Sunday came, they invited their few neighbors to worship God in their home.

My mother gathered wild onions to eat, my grandmother spent a great deal of time gathering buffalo chips for fuel and my grandfather hunted buffalo for meat.

Today I sit comfortably in my modern home looking at the old iron kettle that swung from the back of the prairie schooner. Grandmother cooked buffalo meat in it. It is now a flower pot in my sunny south window. Altho I enjoy modern conveniences, I treasure that old iron kettle.

Mrs. Ivel Curless
R. 6
Wichita, Kan.

Prescribed a Wagon Trip

My mother was an invalid with TB. We lived in western Kansas and, when I was nine years old, the doctors advised my father that a change of climate might help Mother.

Father rigged up two covered wagons. My brother was to drive one, and Father took great pride in making the other one as comfortable as possible for Mother, who had to lie down most of the time.

One autumn morning we bade a sad farewell to an older brother who was to remain behind and set out toward Missouri. I took my little canary, but it soon died from the motion of the wagon.

One night we made camp near an Indian cabin. I had a terrible earache, and we went to the Indian for medical aid. The old Indian blew tobacco smoke in my ear—and it helped!

The next morning after we had started on our way, two men riding horses followed us for awhile and then pulled up beside my father's wagon and began asking questions. They asked to search our wagons. When the search was over, they told us a young girl had been murdered the night before near the place we had camped. She had gone to a spring for water and had been seized and bound to a tree and a knife plunged into her heart. When the officers were convinced we knew nothing about it, we were allowed to go on.

Father bought a farm in Cherokee County, Kansas, and we settled down again. The trip had not helped my mother, and she passed away in April.

Mrs. Susie Bugh
Ewing, Mo.

Recovered His Health

My grandfather was a soldier in the Civil War. He took measles and slept in the snow while he was sick, and this caused him to have lung hemorrhages. The doctors told him that the only hope he could have for recovering his health was to go to Colorado and perhaps the mountain air would heal his lungs.

With his young wife and baby daughter he joined a wagon train for the long and dangerous journey. My grandmother was the only woman, but 21 men made the trip. They drove teams of oxen. Three times during the trip they circled their wagons to form a barricade, put the cattle inside and prepared to shoot it out with the Indians. Fortunately, there were no battles. My grandfather recovered his health and was able to work. The jostle of the wagons during the day would churn butter for them.

Mrs. Anna Rising
Guthrie, Okla.

CHAPTER 6: Outlaws At Large

Grim Hand in Their Wagon

It was mid-October of 1868. The Pendarvis-Roberts-Tanksley caravan was camped for the night along the Marais des Cygnes River near Osawatomie. The Tanksley children, Albert, Allen, Caroline and Melvina, and Junie and Icie Pendarvis raced thru camp and watched the huge flames from the camp fire die away into a bed of red-hot embers. The four families were bound for the Elk River bottoms below Independence, where all had claims.

Margaret Tanksley and Mary Pendarvis, sisters, and Beth Roberts, sister-in-law, were preparing supper. Suddenly out of the twilight, two riders appeared. They showed signs of having ridden far.

"Howdy," the black-bearded one said. "Any chance of gettin' a bite of supper?"

Jim Pendarvis said, "Good evening," in his courteous manner and turned to Bill Tanksley, who was considered head of the caravan.

Bill looked the men over, not liking their rough appearance. "I'll tell the women to throw in some more grub," he said curtly.

Mary Pendarvis carried $3,500 in the pocket of her full skirt. It was the saving she and Henry had made in the ten years of their married life. Henry noticed the strangers eyeing Mary and was resentful, without thought of the money.

Preparations were being made for the night when the bearded

man drawled, "Mind if we share your fire tonight? We'll keep it going."

Tanksley had been hoping fervently they'd go. He turned on his heel, saying shortly, "Pile up anywhere."

As Henry helped his wife and two daughters into his wagon, he looked back to see both strangers still watching Mary. Inside the wagon, he fastened the cover flap securely and placed his pistol under his pillow. He blew out the lantern and they undressed in darkness.

Hours later, he was awakened by Mary's hand pressed against his mouth.

"Someone's searching the wagon!" she whispered. "I felt a hand as it reached thru this side!"

"They're after the money," Henry whispered, realizing the men had spotted the bulky packet in her pocket as she worked around the campfire.

Suddenly Mary's body stiffened and Henry knew the hand now was between the canvas cover and the sideboard. He grabbed, felt the hand, missed catching it. Then he found his gun. The next instant he was out of the wagon, yelling for help as he ran.

He followed the fleeing men, firing as he ran. As his brothers-in-law reached his side, the sound of galloping hoofbeats was heard. They reasoned that further chase would be futile in the darkness. At daybreak they found the blood-covered trail and knew Henry's gun found at least one mark.

The story of the attempted robbery of Aunt Mary has been told and retold to each generation, and each listener breathes a prayer of thankfulness at its happy conclusion.

Charlotte Stark Offen
El Dorado, Kan.

Gun at Hand

My great-grandmother came to Kansas in a covered wagon when she was a young girl. Her family stopped at one of the wayside hotels for a meal.

Great-grandmother was sitting at the table beside an empty chair. There was sudden confusion when two known outlaws came in and one of them seated himself by the frightened girl. She was too scared to eat because the man had placed a gun beside his plate.

When the outlaw noticed that she wasn't eating, he picked up his gun, twirled it around his finger and said, "What's the matter, little girl? Are you afraid of my toothpick?"

Mrs. Jim Crane
6509 S. Broadway
R. 9
Wichita, Kan.

Al Jennings, Outlaw and Gentleman

As a young man Al Jennings lived in Western Kansas, Oklahoma and the Indian Territory. In 1885 his father was county judge at Coldwater. The family was one of much better background and education than most of the frontier people.

Al threw in his lot with the cowboys and it was while he was working in the Cherokee Strip that injustice dealt out under the name of the law took the life of his brother and started Al on the road to crime, he explained in later years.

But neither tragedy nor the role of a bandit robbed Al Jennings of his mark of good breeding. He had his own code of honor and it included loyalty and courage. The settlers were his friends and he never stole from them or missed a chance to do them a good turn.

My father, Henry Durkee, and his wife, Lida, had established a homestead 18 miles south and east of Coldwater. Altho Al Jennings already was a notorious character, my mother never had seen him until one hot day when she was alone with two small children, one a baby in arms. The drouth that year was forcing the ranchers to sell most of their cattle, and my father was away on a cattle drive.

The blistering sun was beating down fiercely upon the plains that August day in the 90s. The short buffalo grass was dry and crackled under the feet of the grazing stock. Heat waves shimmered, and here and there whirlwinds gathered the fine dust and danced away over the prairie.

My mother stood and stared almost unbelievingly at a man making his way on foot across the pasture. She could only speculate on the circumstances that would cause a man to be on foot in this land of great distances.

Men could become ill, have a sunstroke or even lose their reason wandering over plains in the terrific heat. As the fellow approached the lonely little ranch house, fear grew within my mother.

At last, hat in hand, the stranger stood at her door. He was tall, slender, with a frank, kindly expression in his eyes. His black tailored suit, the coat of which he carried on his arm, was sprinkled thickly with dust. The collar of his white shirt was streaked with perspiration. But the smile on his handsome, heat-flushed face was easy and pleasant.

"I have come to borrow a horse, madam," he said. The words struck fresh terror to Mother's heart. There was only one good saddle horse left on the place. To lend it to a stranger—to risk having it ridden down, winded or stolen? He saw her dismay.

"I'll not ride your horse too hard, madam, nor too far. And I will send it back to you safe and sound. But I must have a horse."

She explained her helplessness left alone with two little ones, five miles from the nearest neighbor. Sympathy and understanding were in his intense eyes as he said firmly, "But I must have a horse."

Fighting back tears, Mother pointed out Ribbon, a fleet-footed little saddle mare, grazing not far from the house. The stranger told her that as proof of his word he was leaving her his gold watch which she was to keep until her horse was returned.

The watch was big and enclosed in a solid gold case. She was showing it to the children when the back of the case came open and she saw engraved the name, "Al Jennings."

Next day, scanning the horizon, she caught sight of a rider. Soon her hope for Ribbon's safe return was realized. A man dressed in the worn garb of a settler brought the horse, safe and sound, to her door. Eagerly she reached out for the bridle and patted the mare's sleek coat.

"Mr. Jennings must be a man of his word," she said.

"He sure is, ma'am. He sure is that," the man said.

My mother went to the bureau. In the top bureau drawer lay Al Jennings' watch beside the pearl-handled pistol my father insisted she have at hand, but which she was afraid even to touch.

She gave the watch to Al's emissary. The next day Father rode in, weary from long hours in the saddle and with anxiety in his eyes.

"I hurried as fast as I could," he said. "A train was robbed near Dodge City three days ago. They thought Al Jennings and his gang did it and they were headed this way toward the Oklahoma border."

Then my mother told her story—one that has been repeated many times, and one that never failed to thrill me.

Al Jennings reformed and became a successful evangelist. He later moved to California where he made stage appearances and spoke on radio programs telling of his colorful life in Kansas and Oklahoma.

Ruby Basye
Coats, Kan.

Ran Off Rustlers

My mother drove one covered wagon and my father drove the other when we accompanied cowboys driving a herd of cattle thru the most dangerous sections of the Nebraska sand hills. I was a little girl, but I remember it well.

We had been advised not to drive thru this section with the cattle because rustlers might attack. But my father said it would take too long to drive around—so we drove thru anyway.

One night we were attacked by rustlers and they tried to

stampede the cattle and almost did. I remember the cattle running and bawling. My mother and I ran to the wagon and crept in for protection. It was lucky we did because our tent was torn down. The rustlers did not get the cattle, but we discovered the next morning that they had driven our hobbled teams away.

The cowboys found the horses the next morning, and we finally reached our destination in safety. Guess we had too many cowboys for those rustlers.

Mrs. Roy Kirkwood
Turney, Mo.

Border Warfare

When the border warfare between Kansas and Missouri broke out, it became necessary for my Grandfather and Grandmother to leave their home because Grandfather had spoken his views too freely and had made enemies. In fact, he had to remain hidden to save his life.

One night they put a cover on the old wagon, loaded their meager possessions and headed west. Grandpa went ahead on foot and Grandma and the babies followed in the ox-drawn wagon. Grandpa, in scouting ahead, came upon two men burying a man they had killed—all neighbors, but bitter enemies because of the slavery question. Grandpa stopped behind a tree and watched. He brought out his old muzzle-loader and prepared to kill the men who had killed his friend.

Just then he heard the ox wagon coming, bringing Grandma and the babies. The realization struck him that he was all they had to see them to safety. So he let the hammer of the gun down and slipped quietly away. I don't suppose those men ever knew how close they were to death or that they were saved by the sound of an ox wagon.

Mrs. Raymond Williams
R. 4
Paola, Kan.

Lawless Brothers

My mother's half-sister, Aunt Ellen, lived in the part of Oklahoma that was then called the Indian Territory. She had married a good man, but he had two lawless brothers in trouble about the theft of some cattle.

Every fruit tree on Aunt Ellen's farm that year was heavy with fruit. She wrote my mother and told her to come down from Missouri and bring her jars and fill them with fruit. I was eight years old and was taken along to care for my year-old brother while Mother and Aunt Ellen canned.

We made the trip in a covered wagon pulled by two huge mules. On the second day there, I saw a man skulking furtively thru the orchard with a gun in his hand. I ran to the house and told my uncle a man was going to steal his fruit. My uncle just looked sad and told me not to go to the orchard alone. The man was his brother and another brother was asleep in the barn loft.

That night the two brothers came to the kitchen door, stood their guns against the door jamb and said to my 18-year-old cousin, "Hey, Alice, have you got anything to eat?"

They were her uncles and she was used to them, but I was almost petrified at the sight of the guns. Alice set out some food, all the time telling them off at a great rate. "What do you mean, hanging around decent people? If you want to live like devils, you stay in your own hell! Don't be hiding around here and getting us into trouble. Pa and Ma are good and they are ashamed of you."

The two men seemed amused at her spunk in telling them off. They ate, picked up their guns and said they were sleeping in the barn loft.

Aunt Ellen had forgotten to bring in the peaches that had been drying on racks in the sun and she was afraid it would rain in the night. She asked me to go out and help her bring them in.

As we went Aunt Ellen talked quite loud. I said, "Not so loud, Aunt Ellen, or those men in the barn will hear you."

"That's why I'm talking so loud," she whispered. "I want them to know it's us and not officers—so they won't shoot." Then

I really was scared.

It was a night of terror for me and I was glad to return to my peaceful home.

Daisy M. Hyde
34 Ambrose Ave.
Pittsburg, Calif.

CHAPTER 7: Treacherous Torrents

Crossed Missouri on Ice

We crossed the Illinois River on a flat boat with no sides on it when we traveled from Illinois to Kansas in 1879. The boat was powered by two donkeys going around and around as if they were grinding cane.

The boat across the Missouri River had a coal-burning engine and was not quite so frail looking. We saw wagons going west with the slogan, "Kansas or Bust" and we met wagons headed East with the word, "Busted."

When I was a bride in 1884, we crossed the Missouri River on the ice. We had been to Missouri in our covered wagon, but we were buying a farm in Kansas and had to get back to fill out some papers. We had a small stove in the back of the wagon because it was very cold and there was snow on the ground.

One day was so cold the wagon squalled all day and we made only eight miles. When we finally got to the Missouri it was frozen solid, and my husband decided to try to make it across by driving over the ice. We stopped in a little burg and bought some rope and my husband hired two men to help us. We tied the rope in the end of the wagon tongue and hitched one horse to it. One of the men led the horse and I led the other horse.

The other man walked along the rope and was to cut it quickly if the wagon started to sink. That way at least the horse would have a chance. My husband carried the trunk so we would have some clothes if the wagon went down.

We made it all right. I haven't seen the Missouri River since—and I don't want to!

Mrs. N. E. Cannon
Butler, Okla.

Quicksand Danger

We packed our belongings into two covered wagons and a spring wagon to travel to Texas when the sand and dry summers drove us out of Kansas. The spring wagon was for the ladies and for us little girls. My father's brother and his family went with us.

We were six weeks on the road and Mother got very tired baking biscuits in a skillet over a campfire for so many hungry people. But Mother could bake the best biscuits in the world with half a chance.

People who knew the streams along the way warned us always to water our horses before we started across. If the animals stopped to drink, they might sink fast in the quicksand. Many a wagon had gone down! There were no bridges, of course.

Two of our teams had made it thru one stream when the third team began to struggle. Frantically my father called to my uncle to bring back the lead team. Father carried us children to safety on the opposite bank. The men were planning to unload the wagon when the lead team managed to pull the sinking team to firmer footing and they made it across all right. My heart beats an extra pitter-patter when I think of it even today.

Mrs. Viola Hall
Altamont, Kan.

Drowned Her Oxen

A widow with seven children, ranging in age from 21 years to three years, left Texas by ox team on May 14, 1867, headed for Kansas and her relatives. They loaded the farm implements and household goods in one wagon and the bedding and clothing in another. The oldest boy to help with the driving was 14.

Red River was up and they had to camp three weeks and wait for the ferry to run. When they got to the middle of the river, the ferryman got panicky, said they were loaded too heavily, and pushed one wagon and team into the river! The oxen were yoked and they drowned. The widow lost all her farm implements and household goods.

They ran out of money and had to stop and work to make enough to come on to Kansas. The two oldest girls worked for a dollar a week and the boy got work with the team. It took from May until August for them to reach their relatives. The widow was my great-grandmother.

Mrs. P. E. Dieffenbaugh
1230 Grant
Clay Center, Kan.

Baby Dumped Into River

I was born on a homestead in the Oklahoma Territory in a one-room shack with cloth tacked over the one window and a blanket over the door. The wolves were so thick my mother said she was afraid to go to sleep at night for fear they would carry off the children.

When I was six years old, we started out in two covered wagons to go to Arkansas for my mother's health. There were seven children living. We had two wagons and were driving several cattle. Father drove one wagon and Mother drove the other. We had a big wood-burning cookstove in Mother's wagon to keep us warm and to cook on. We pulled into a creek, and Mother's wagon slid up on a big rock and turned about half way over. The cold water on that red-hot stove broke it. Mother, one little boy and the small baby were dumped into the icy water. But they all made it out and dried out in a house near the river. We lost quite a bit of our stuff, tho.

Mary King
Adair, Okla

Up to Their Ears

We smaller children drove the cows behind the wagons when we came to Kansas by covered wagon—and we had the time of our lives. When we got to one river, it was out of its banks. I think it was the Neosho. We had to lay over two days, but on the third day my father said we had wasted enough time. He put the family all in one wagon, tied one team to the back and tied the cows to the sides. Father drove and we started across.

Very soon the water was in the wagon bed. All of us climbed on top of the bunks and held as many things out of the water as we could. About half way across the river, both horses and cows had to begin swimming. All we could see of the horses were their ears bobbing up and down! By some miracle, those animals swam us across safely.

Mrs. J. G. Sitz
WaKeeney, Kan.

Lost Their Money

On our way to a homestead in western Oklahoma, our heavily-loaded covered wagon started across a creek after an all-night rain. The crossing was very steep and slick and we had only one team. Just as we got ready to pull out, a coupling pole broke.

Papa was pulled out of the wagon, and Mother and we children were left in the wagon with the water rolling all around us! People were kind and helped us unload and carry our things ashore.

Two days later we lost all the money we had—$50. Papa had had Mama carry it in her handbag because he was afraid someone might rob him. My baby sister had been playing with the handbag and we decided she had thrown it out of the wagon.

We camped for three days, hoping someone would come by who had found the money or had heard of someone who had. Papa offered a five-dollar reward for the return of our "fortune." Finally, an elderly man said his daughter had found the handbag and everything in it.

When we asked her, she denied finding it. But after her father talked to her, she gave it back to us and took the five-dollar reward.

We were 21 days on the road and we walked a good bit to make the load lighter. The shoes that were new when we left Grandfather's house were worn out when we got to our homestead.

A few days before Christmas we began our new life in a dugout. It was a harder life than we'd ever known, but I'm sure it was not as hard for us children as it was for our parents.

As soon as we put out the coal-oil light at night the centipedes began to crawl. Our $50 was gone and we all felt so afraid about the future. There was very little to live on—mostly cornbread and a drink for breakfast made from burned cornmeal.

The fall rains were heavy and incessant. Our dugout started to cave in and loads of dirt fell on our beds and nearly buried us girls alive. Our parents had just dug us out when an awful amount of dirt fell and our dugout was ruined. But we all escaped alive.

Mrs. Grover Harden
Fay, Okla

Turbulent Missouri

Wheat that we ground in a hand-turned coffee mill was our chief source of nourishment when we six girls, our two brothers and our parents traveled by covered wagon.

A neighbor who received a Civil War pension was in the wagon train with us and he used his money to help us in all our needs. He had only one son himself. I was very fond of these neighbors and I chose to ride with them instead of with my own family.

I regretted my choice, tho, when we had to ford the big Missouri River because the Civil War veteran decided that his wagon should lead the way. It was a time of turbulent streams, and my father and the neighbor talked a long time about whether it was safe to cross or not before they started.

I shall never forget that experience. I felt we never could make it. The horses often swayed downstream with the current, and several times we almost overturned. But God preserved our path to dry land.

When we got to our new home we were delighted to find a lovely orchard of big, red apples, all ready to eat, and a house large enough for all of us and with a wonderful fireplace—something none of us had seen before. Friendship and dependence on Divine Power were great assets to the early pioneers.

Mrs. Harvey J. Brown
Williamsburg, N. M.

Thru the Overflow

On a bright, pure morning long ago my father and mother, we five boys and our baby sister started to Nebraska from Iowa in a covered wagon.

We had not gone far until we came to a river that had overflowed its banks in a big way. We made it across the bridge all right, but on the other side there stretched out a great expanse of water. I remember I thought it looked like the ocean must look.

My father got a long stick for a cane and led the team feeling ahead each foot of the way. Dry ground was fully half a mile beyond. I remember how scared I was and what a relief we all felt when our wagon pulled out of the water at last.

C. A. Strawn
Parker, Wash.

Time of High Water

My father made the bows for the two wagons we used on our trip from Colorado to Iowa, and I sewed the covers out of eight-ounce duck. We fastened strips of white oilcloth along the tops of the covers so we would be sure to have dry beds in our wagons.

Father made what was called an overjet for each wagon so our bedsprings would fit nicely. We packed dishes, clothing and other things in boxes and stored them under the beds.

If there was a cloud coming up at night when we camped, Father would not go to bed with the rest of us. He walked the road and watched and if he thought there was danger of a tornado, he would wake us and hurry us to the next homestead in hopes of finding a storm cellar. We were thankful we never were in the heart of a storm, tho we would drive thru torn-up sections the next day and see what we had been lucky enough to escape.

When we got to the Missouri River we found it was the highest it had been in 40 years! We had to camp for two weeks at Portsmouth, Neb., waiting for the water to go down. Finally, it went down enough for us and our wagons to get across on the cable ferry.

Mrs. H. M. Groves
Cainsville, Mo.

Pioneer Child's Fright

I was only three years old when we traveled by covered wagon from Missouri to western Kansas, but I remember how scared our bay mare, Queen, was when we were near the few trains we saw. She would squat and tremble all over.

My father took a claim 18 miles southwest of Scott City. We lived there three and a half years before we starved out and started back to Missouri—in a covered wagon again.

We took a western pony back with us and I remember that in St. Joe the streets were so narrow she walked on the sidewalk some.

When we got to the big Missouri, we had to cross it over a railroad bridge. There was water swirling below which was frightening enough—and then I looked back and saw a train coming behind us! Tho we got across all right, I've never forgotten how frightened I was.

Mrs. Florence Hoover
Springfield, Colo.

CHAPTER 8: Hopes Were High

Land of Milk and Honey

How thrilled my parents and we two little girls were as we loaded our covered wagon to go from Iowa to a homestead in the Indian Territory! We were going to a land of milk and honey, my father said. I was only four years old, but I remember the sense of pleasant anticipation we all felt.

The things my parents and grandparents loaded into that wagon! The shelves that were part of the overjet were stacked with boxes of dried fruit, sacks of navy beans and cans of honey, all from Grandfather's farm. They built a bin under the spring seat and filled it with potatoes, and Father built a chuckbox across the back of the wagon and feed boxes for the horses. A lantern was hung on one wagon rod and an iron teakettle on another, both on the outside of the wagon at the back.

In spite of having arranged and rearranged, a lot of things were left on Grandfather's porch. I was told that I absolutely could not take my precious kitty, but while my young parents were getting some last minute advice I stuck the kitten in the iron teakettle.

As we traveled everything was new to us—and so exciting. It seemed everyone was going to the Territory. Several covered wagons with big families usually camped together at night. The men would gather around the fire and talk of horses, and the women would talk of the new land where they were going.

The orchards along the way were hanging full of fruit and the fields were overflowing with shocked feed. Everyone seemed so generous and told us to help ourselves and wished us well on our journey.

Then somewhere in the southern part of Kansas we ran out of nice farms and the country became wilder. There were no bridges across the rivers. In one stream over went wagon, kids and all! But the potatoes were the only total loss we had. Were we ever soaked! We spread the soaked beans and fruit out to dry along with all our goods.

Not until the next morning did I remember my kitten and run to let her out. I was heartbroken to find that she was dead. My mother was crying and I thought it was because my kitty was dead. I told her not to cry because I'd get another when we got to the land of milk and honey. Later I learned that she had cried because she thought her new Singer sewing machine was ruined.

Fannie Brollier
Pond Creek, Okla

Pa Dreamed of Kansas

Our neighbor in Iowa made a trip to Kansas to see the country when I was seven years old. On the trip he captured a buffalo calf and brought it home with him.

My father went to see this neighbor and got a good report on Kansas. When he saw the buffalo calf, he caught an awful Kansas fever! He was so eager to go to the new land he talked of nothing else.

I heard my mother tell Pa that if they could pay all their debts, have a team, wagon, harness and $500, she would go with him to Kansas and grow up with the country. This was about 1875, and it was 1877 before we met Mother's specifications so we could go. I was nine years old.

When we came over the hill above Burr Oak, where we settled, we could see the little town nestled in the bend of White Rock Creek. There were a few log cabins, one general store, a post office, a blacksmith shop and a saloon. No sidewalks, just a path from one store to another.

Game was plentiful with prairie chickens, quail and some deer. There was plenty of wild food—wild plums, Indian breadroot, buffalo peas and wild onions. I am 87 years old.

Clark Foster
Burr Oak, Kan.

Some Hearts Were Light (Reprinted from Capper's Weekly)

Did we have fun in pioneer days? It depended on the ability of each individual to get joy out of the most trivial things and to sustain his spirit of great adventure in the midst of unglamorous surroundings.

But I remember many things in which we rejoiced and found much pleasure. The once-a-week mail added great zest on the day it came. And we had a week of anticipation and speculation before it arrived.

There were the days we spied a rig coming down the long road toward our home. We wondered who our callers could be and we had ample time to "tidy up" and be in the yard to welcome them with broad smiles and real western hospitality.

I found pleasure in helping gather the fuel for winter. We had tired bones, of course, and such sound sleep at night! Day after day we hauled in cow chips and stacked them in neat piles. In time we became experts in picking the right kind as to age and color so that we had the hottest flames with the least ashes. There was a deep feeling of satisfaction when we pronounced our stack the best and biggest in the neighborhood.

And those majestic mirages that delighted and entertained us on rare mornings and days! On those magic mornings houses appeared where they'd never been before. Distant buildings and towns seemed to have come into our neighborhood. On warm, sunny days there were elusive bodies of water playing about the prairies and horses and cows looked as if they were walking on high stilts in the shimmer. No amount of money could have purchased such glorious displays of nature's grand phenomena.

I admit there was worry, hardship and heartbreak in pioneer days, but there was light, color and joy for hearts capable of experiencing these feelings.

Mrs. Louise Brumfield
Jetmore, Kan.

Joy of Fresh Experiences

Pioneers enjoyed life. For one thing, most of them were young people or in the prime of life. Everything was new to them and there was so much to see and do. Their dugouts and sod houses probably meant as much to them as our houses do to us now. They didn't expect much, and so it didn't take much to make them happy.

People didn't hurry like they do now and if two people met on the road they stopped and talked awhile whether neighbors or strangers. School houses were social centers and places for religious services, literaries, political debates, box suppers, ciphering and spelling matches.

Near election time there were big political meetings with barbecues. People came from miles around to spend the day and eat and laugh and joke and forget about the grimness of prairie fires, blizzards and grasshopper plagues.

Dora Bucklin
Orleans, Neb.

Around the Campfire

Twenty-seven of us—five families in five wagons and two old buggies traveled to our claims in what is now the Panhandle of Oklahoma. It was grass as far as you could see.

We made quite a train and had a good time together. When we camped at night we left our campfire burning after supper. The children played games in the twilight and we grownups sat around the fire and talked and joked. I look back now and see how happy we were. We didn't know then the hardships that were ahead of us.

Mrs. A. H. Turner
Elkhart, Kan.

Prairie Adventurer

Grandpa Linzy had a bright yellow beard, eyes that flashed blue fire, a Viking ancestry and a boyish enthusiasm for pioneering that did not dim until he was well past 60 years old. Things always were going to be better on some wonderful new frontier! It didn't matter too much that they weren't because by that time Grandpa already had started dreaming of the next big adventure.

So far as I know, Grandmother Mary never rebelled. It was Indiana to Iowa, to eastern Kansas, back to Iowa, down to Pottawatomie County, Kansas, away out to Ness County, Kansas, and then back to Pottawatomie County again—and every mile of it in a covered wagon.

Grandmother was a mild-mannered, gentle woman whose first husband had died the night her first baby was born. She was much beloved by all of her relatives and neighbors, and the only criticism of her I ever heard was that she was a little more partial to boys than to girls. She adored her Linzy and was proud of his bigness.

Many a woman of her day put her foot down and refused to let her man pull up stakes too often. Tho Grandmother well knew that her husband's glowing predictions for better days ahead weren't likely to come true, I think she secretly enjoyed their adventures as much as he. At any rate, she bore her seven children in various dugouts, soddies and shacks, and sometimes waited only until the next baby was born to start out in the covered wagon again. My mother was only six weeks old when they jolted to Iowa's "greener pastures." Once they did stay in one place long enough to have a good house, one built of quarried stone—out in Ness County.

When they first homesteaded in Pottawatomie County, they had to go to Fort Leavenworth for supplies. Grandpa and my two half-grown uncles made the trip, a long day's journey each way. Once they were to buy supplies for neighboring homesteaders, too, and Grandpa carried a money belt with a good bit of cash in it. For some reason, they found themselves many miles from the Fort when night fell.

They made camp and as they cooked supper three rough-looking men, wearing guns, rode up and asked if they could camp there, too. Grandpa consented and decided he would "draw them out" before they went to bed and see if he would feel safe going to sleep with his neighbors' entire cash reserve on his person. The men wouldn't "draw out." They sat silent and uncommunicative, tho Grandpa was a great talker and a good storyteller and usually was successful in sparking off a lively conversation. That's when he and the boys decided they would take turns keeping watch thru the night for foul play.

They had no guns; an ax was their only weapon. As the night deepened and they realized their helplessness against guns on a lonely plain, father and sons felt more and more uneasy. Every time a horse would snort or one of the strangers would stir in his sleep, their hearts raced in panic. Tho they had planned for only one person to watch while two slept, none of the three slept a wink all night, tho my younger uncle, 14, is said to have dozed off at daylight. Nothing happened, and the strangers, refreshed by sleep, were much more talkative in the morning than Grandpa himself. Grandma said that when her men got home from that trip they were so tired they were grouchy as three bears.

Once the family had been eating cornbread for months because the weather was too bad for a trip to the Fort to get white flour. Grandma could make fine, fragrant light bread—and the family longed for a loaf of it in a way we, who have it every day, cannot imagine.

At last the weather "broke" and Grandpa and the boys made the trip. They left Grandma's list with the grocer and went out on the streets to catch up on the latest news.

Grandma had fresh butter churned and was all set to do a baking the minute the menfolks got back with the flour. The little children awaited the warm white bread as eagerly as if it had been candy. Alas, when the wagon was unloaded, there was a barrel of cornmeal instead of a barrel of white flour!

For once, Grandma was a little cross. "If you'd seen to the stuff yourself, Pa, this wouldn't have happened! But, no, you couldn't wait to get out on the street and start chinin'! "

My mother, who was a little girl then, said she noticed, tho, that Grandma wasn't content until Grandpa had told her every last scrap of the news he'd garnered!

Alta Maxwell Huff
Topeka, Kan.

CHAPTER 9: Children Loved It

Tumbleweed Tree

When we lived in a two-room house—half dugout and half sod—on the Western Kansas prairie, we had no toys. We made our own amusements. We learned early to ride our ponies, and in the spring we gathered the lovely wild flowers.

Well do I remember our first Christmas tree. It was a tumbleweed decorated with paper chains and pictures taken from packages of Arbuckle coffee. My brothers learned to braid different colors of hair from the horses' tails to make attractive belts, quirts and bridles. Little sister spent much time holding the strands as her brothers braided.

The prairie was plentifully strewn with dried buffalo horns. We scraped and polished these to a lovely shining black and made coat and hat racks from them. We were happy. That fact stands out above all.

Mrs. J. W. Edwards
Meade, Kan.

Pure Pleasure

In March, 1882, when I was eight years old, our family traveled by covered wagon from Lincoln, Neb., to near Mitchell, Dakota Territory.

We—a girl cousin, my two sisters and I—reveled in the

delights of the journey. We searched the roadside for treasures; we eagerly blocked the wagons, three of them, with rocks as the horses rested up the long hills. We ran ahead as we neared the top for a view of the road ahead, the next hill top, smoke from a town or the Missouri River bluffs.

At night we brought the tent pins and watched the tent go up. We gathered sticks and cobs for fuel for the stove. Then how good the re-heated meat, hot biscuits and jam tasted!

From a nearby strawstack we got straw on which to make our beds and slept well, sometimes with coats and hoods on. Not even a cold troubled any of us.

At Niobrara the ferry acoss the Missouri River was exciting. Pet and Prince trembled at the first chug of the engine but Father, with his hand on each nose, said, "It's all right" and neither one moved a foot.

All the journey brought daily thrills for three weeks with health and hope, which ever since have been a joyful memory.

Miss Mabel Gibbs
Canon City, Colo.

Schooner Thrill (Reprinted from Capper's Weekly)

In the autumn of 1879 when I was nine years old, President Hayes came to the county fair held in a town 30 miles from our Kansas homestead. Everyone was excited and Father began to fix up the old prairie schooner in which we had come from Iowa to Kansas seven years before. We decided to go a day early and camp on the grounds over night. Mother prepared a basket of food and extra clothes for the big day. We could hardly wait to get started.

What a grand spectacle for our young eyes! A coach drawn by four plumed black horses appeared. In the front seat sat President and Mrs. Hayes. The people shouted and whistled and waved and the President lifted his tall silk hat and bowed right and left.

Later, a big dinner was served and we children hurried to the dinner scene to get a closer view of the President. We were awed and bashful and stood afar off. Then, oh joy!—he saw us and spoke in a kindly way and shook hands with each of us.

After 70 years, I still recapture some of the thrills of that day. When we boarded the schooner for our trip back home, our heads were in the stars. Life could never again be dull. We had seen the President of the United States!

A. M. Heistand
Iola, Kan.

Pig-a-Back Pioneers

We saw our first railroad when I came with my parents in a wagon train from Texas. There was whooping and cheering from the men and the boys—and all of us had to take a walk on the ties.

When we camped at night, the older children taught us little ones how to make shadow pictures on the tent walls. I was only three or four, but someone taught me to make a bunny rabbit. When we came to streams, the men and big boys would carry us smaller ones across on their backs to save the teams and because we thought it was fun. I have no idea how long we were on the road and there is no one left to ask.

I remember that when we reached our homestead, we ate off a wash tub turned upside down, and the grass around the place was so high that we children could hide from each other in it. Our first school was three miles away, and the men dragged a big log thru the tall grass to make a trail so we wouldn't get lost.

Mrs. E. Mathis
Custer City, Okla

Prairie Dogs Amused Them

There were eight of us children and Father and Mother when we set out from Missouri in two wagons with overjets and wagon sheets over the bows. We left early and made 25 miles the first day.

When we knew we were coming to a town my sister, Irene, and I would stick our heads out the round hole in the back and wave to the boys and girls. We thought this great fun. At night we would build a campfire and we children would play and enjoy it so much.

We stopped at Grandfather's place and he gave us ten cows. Then we children had to walk and drive them. When we got into Oklahoma we saw prairie dog towns. We thought it was hilarious to run hard toward the town and watch the alert little dogs sit on top of their holes until we were almost on them and then flip down into the holes.

Mabel Newberry
Bovina, Texas

Rich Childhood

Imagine taking a family with small children out to a claim absolutely bare of improvements—and in mid-winter! That is what my parents did, but I'll always be glad for my pioneer experience. There were five of us children, from 14 years old down to the baby. I was seven.

We lived in a tent until Papa built a one-room stable in which we lived until he built a small house. The trips to town were made only once a month. Papa would start long before daylight, and all day long we were excited and full of expectancy about the things he would bring home. At dark we would start listening for the wagon and when we did hear it we would run down the trail to meet him. Flour, dry beans, coffee, sirup, bacon and dried fruit made us feel rich.

When spring came, what fun we had planting the seeds we brought from Kansas! The sod plow turned long, black ribbons, and how the plants did grow in that virgin soil! No need for cultivation that first year. We just dug a hole and dropped in the seed.

No children in any time could have had more fun with their games than we had on our homestead. We played daily with a family of dolls made from rags and paper. Wagons made from oat boxes and drawn by cucumber or wild gourd horses hauled loads of hay (grass) to miniature barns. Fences were sticks driven into the ground and strung with string. Once a tornado (whirlwind) upset the whole doll family. Some were blown several "miles" and others hurried to caves we had dug. One day a giant bird (chicken) grabbed a man doll by the leg and ran away. Another day while fishing in the lake (a shallow pool below the spring) a great fish (crawdad) tipped over the tiny boat and the whole doll family fell in the water.

Once I secretly slipped a newly-cut-out baby into the doll bed. My little brothers were so surprised to see an addition to the family. "Just like grownups," I explained in a superior way. "You just wake up some morning and there's a new baby."

Later that summer other families began moving into the new country and we had neighbors. Papa was a carpenter and built several houses for newcomers. He was walking home at night once thru shoulder-high blue stem when a panther screamed near him. The only weapon he had was a pocket knife. He opened it in readiness, but did not hear the panther again. Coyotes were so thick they often chased our dog right up to the house. I could write a book about my happy and exciting pioneer childhood.

Mrs. S. H. Cowl
Stratford, Okla.

Lonesome Prairie Music

When I was quite small my father rigged up a covered wagon and we went to Texas for my mother's health. Some friends were moving to western Kansas at the same time, so they packed their belongings in a wagon and went with us.

They had a small organ tied to the back. Every night when we camped they took it off and someone played it. I'm sure there never was any sweeter music than that organ music played under the stars on the lonesome prairie.

My two brothers slept on the floor of the wagon under the bed where my father, mother and my sister and I slept. It was like crawling into a cave to get into their sleeping spot. In the daytime I used to wish I was a boy so I could sleep in that wonderful cave-like place, but when night came I was glad to be a girl.

Our covered wagon trip was a pleasant subject in our home for many years. I was the youngest in the family, but I must have enjoyed it most. Altho they all enjoyed talking about it, I was the only one who longed to go again.

Of course, I didn't have to worry about such things as finding a place in an already overcrowded wagon for our dog, whose feet were so blistered by the hot sands he could hardly walk, or about finding food for the family in a strange country where houses were few and far between. Childhood is a wonderful time for adventure!

Dorothy Hiers
R. 2
Council Bluffs, Iowa

Indian Holiday

A six-year-old girl sat out in the hot sun and blistering sand of brand-new Oklahoma and watched the covered wagons go by. Pioneers? Homesteaders? No. They were Indians on their way to the Indian Fairs and Dances.

Between the lumbering old wagons were young braves on their pinto ponies, and there were dogs of every description. The Indians were eager and excited. They were going to the Fair!

My parents and I rode over to the festivities once. Don't tell me Indians are silent people! There was plenty of chatter and fun. Groups of young girls, with gay ribbons entwined in their long black hair and many strings of bright beads around their necks, cast roguish glances at the young braves grooming their horses for the bow and arrow contest. Admiring glances found their way back to the girls.

When darkness fell, many bonfires were lighted. It was beautiful and enchanting to me. At a late hour a huge fire was built. The chief, resplendent in his huge headgear, rode into the bright light on his speckled pony. He raised one arm high. This was the signal for the dance to begin. How colorful when they all circled the fire, dancing and chanting! Some fell out and others joined in. It went on and on.

A week later I watched the covered wagons going home. Even the dogs seemed to be drooping. Canvas flaps were raised and beady black eyes stared at me, the little paleface, as I watched them go by.

Mrs. Herman Hertzberg
South Main St.
Eldorado Springs, Mo.

CHAPTER 10: Hospitality Was Warm

No Wayfarer Turned Away

In the 80s many covered wagons passed our place. Some homesteaders camped in our yard over night. In the spring they went west. Some had signs painted on the canvas covers like "Kansas or Bust," "Going to the Promised Land," and "Home, Sweet Home."

In the fall many wagons came back and the signs had been changed to "Busted, by Gosh," "Promised Land Was a Mirage," and "Coming Back in the Spring." The people going west were cheerful and hopeful. Those going east were ragged, gaunt and tired.

One evening we saw a covered wagon coming up the road. A man and a horse were pulling it. At first we could hardly believe our eyes, but as they came nearer we saw it was really so. The man explained that one of his horses had died on the road. In the wagon were a woman and three small children. They stayed all night with us, and the next morning Father gave them our old family horse. They were so grateful they cried. But after they were gone it was I who cried. The horse had been my pet.

Another time a wagon with a horse and a cow hitched together stopped at our place. A man, woman and baby were in the wagon. They had taken the cow along so the baby would have milk. The man tried to joke about his team, but the woman cried. This time we had no horse to give away, but a neighbor traded a horse for the cow and gave them $5 to buy milk. It is doubtful that they had to use the money for milk because in those days people were generous with what they had and shared with one another. No wayfarer was turned away hungry. If money was offered for food and lodging, it was declined with the words: "You can return the favor by helping someone else in need."

As I look back at the covered wagon days I marvel at the courage, resourcefulness and helpfulness of the early settlers.

Dora Bucklin
Orleans, Neb.

Some Paid—Some Didn't

My mother was left a widow when she was 46. She had four girls, the eldest was 15 and I was the youngest—five. Near us was an ideal place for homesteaders to camp—plenty of water for their horses and timber for a windbreak.

When they came to Mother's to get eggs, milk and chickens, some would offer to pay and others had tales of misfortune. Mother was kind and willing to help anyone in need.

One evening when some folks came by, Mother was just a little afraid because it was quite late. A pioneer mother came to the door and said her son had broken his leg and she was afraid for him to sleep in the damp wagon. She asked if she could bring him in our house for the night. Mother let her bring the boy and stay. They were quite nice and seemed to appreciate it so much. We were not so doubtful of people in those days as we are now.

Mrs. E. G. Estes
Allen, Kan.

Christmas With Strangers

Two large trees in the road near our home made a favorite camping place for homesteaders on their way to the West. When someone announced, "Covered wagon coming!" Father would saunter out to the road for a chat, and he brought back many interesting tales as to where the travelers came from and where they were going.

I well remember one Christmas. The temperature had been dropping all day, and by night it was bitter cold. Father came in and said that there was a covered wagon outside and the people wanted shelter for the night. Mother agreed and they came in, seven of them. They brought their bedding and spread it on the sitting room floor.

I used to think that to travel in a covered wagon would be the most wonderful thing that could happen to me.

Mrs. Lillie K. Sprouse
Braymer, Mo.

Fireplace Blessings

We moved in a covered wagon with a team of mules into the Ozark Mountains in Missouri when I was 11 years old. This was long before the day of good roads and we were able to travel only about 25 miles a day over the rough, rocky, muddy roads. The hills were very steep.

When the sun went down we began to watch for a farm where we could spend the night. Each night we found a friendly family to take us in. They had no spare rooms or beds, but they let us spread our featherbeds in front of their fireplaces. The month was December, and this kindness was a real blessing.

First we built a campfire in the farm yards and Mother cooked our supper. In each instance, the settlers would allow her to bake a large pan of biscuits in their cookstoves. Each morning Mother prepared enough food for noon so we would not have to stop and cook during the day.

Mrs. J. E. Bruce
Medora, Kan.

Gave Them a Team

It was a time of hot, sultry days when our family of eight

moved from Iowa to Kansas. The fruit lay on the ground, apples rotting. We all took terribly sick. We had malaria fever and chills. In a short time we had to give up our oldest sister in death when her malaria was complicated with measles. It broke our hearts, and the rest of us were still sick.

"You folks are from up north where it is cooler, and if you stay here you will die," a doctor told us. How could we get out? No money and all of us so sick. A cyclone had killed our team.

The neighbors gave us a team of horses and must have given us some money, too. We all climbed into our covered wagon, taking our bottles of medicine with us. The doctor told us to go west. After a few days, we began to throw our medicine bottles away! It was wonderful to feel well again! We went 350 miles west and some north. All of us recovered quickly except our brother, John, who had malaria for a year.

Mrs. Eva Teeters
1331 S. Osage
Independence, Mo.

Rest for the Weary

The rigors of the new, unfriendly country—drouth, blizzards, lack of resources and, finally, starvation—drove some of the settlers back to the wife's folks. Many of these stragglers stopped at our home on their way back East.

My parents—bless their memory!—always took them in. Mother fed them at a wellspread table, and father fed their teams from his meager supply of corn and oats. After a few days rest, they went on their way refreshed.

The land these settlers were forced to abandon is now an empire of wheat and oil. But normal human beings can't live on hot winds, blizzards and prairie grass.

Mrs. Pruda B. Utley
407 S. Sixth St.
Arkansas City, Kan.

School by the Trail

I am 96 years old, and about 86 years ago I was a pupil in a small country school in western Illinois. At that time covered wagons were passing on their way to the bridge over the Mississippi. They were on their way down to Kansas to settle.

In one wagon I remember was a man and wife and three children. They had a fine team, a cow tied on behind and a dozen chickens in a box. On one side of the canvas was printed in large letters, "In God We Trust." On the other side were the words, "Kansas or Bust."

Two or three months later I saw that wagon return. It was minus the cow and chickens. On one side was printed, "In God We Trusted," and on the other side were the words, "In Kansas We Busted."

Mrs. N. C. Bowen
Council Bluffs, Iowa

Claim Shack Parties (Reprinted from Capper's Weekly)

The first winter we homesteaded in western Dakota, the snow came early and it was very cold. But we didn't stay home! We hitched old Bess and Lady to the sleigh or the democrat and wrapped up in robes and hied happily away to someone's claim shack to dance all night to the music of a mouth organ or any other instrument that was handy.

The hosts usually piled their furniture out in the yard for the occasion, all but the stove. We had to have warmth—and hot coffee to go with the sandwiches and cakes we had brought.

Everyone knew everyone else and we borrowed from each other such things as kerosene for our lamps and medicine for the ill. We helped each other with the harder tasks and laughed together and shared ourselves and our provisions as people are intended to do.

Once when a blizzard hit, my sister and I were batching in a claim shack. We came home from school almost frozen to find that two men caught out with loads of wood had taken shelter in our shack. I can thank one of them for saving my ear. It was frozen stiff and stuck to my scarf. We two girls would have been afraid there alone, but the men kept us visiting and got in the water for us and kept the stove red hot.

One old pioneer woman we knew told of setting a trap for a mouse. She heard a clatter one day and was surprised to see a rattlesnake appear on the floor with the mousetrap on his head! She ran for her son, who came with a hoe and killed it. The whole community rocked with laughter at the idea of a snake weaving around with a mousetrap for headgear.

When we went to dances, we danced all night. There were no cliques and we danced with all, young and old alike. We went to have fun and we had it. On Sundays we went to various homes to sing and play games and pranks on each other.

We were poor, but we enjoyed life.

Mrs. Ray O'Connell
Milesville, S. D.

Homestead Prayer Meeting

I am 83 years old and I believe people enjoyed themselves as much in pioneer days as they do today. In my early girlhood my father would hitch a span of mules to a wagon box on "bob-sleds" and put hay thick in the wagon bed. He would drive to the home of neighbor after neighbor and gather up old and young alike. We all went to someone's home where we had prayer meeting, play parties and spelling matches.

I distinctly remember one prayer meeting. The host brought in sideboards from his wagon, placed one end on the bed and the other on the cedar churn. The other sideboard rested on the bed and a keg of kraut. I had to sit over the churn and I could smell that old buttermilk. My chum sat across the room from me and kept sniffing like she smelled something bad. It made us giggle, and when we compared notes later we discovered that she was sitting over the kraut.

But how fervent and sincere were the prayers offered to the Lord. Our neighbors were all clean characters. We had wonderful parties, too. We played Skip to My Lou, Happy Is the Miller Boy and My Ship Has Arrived.

If a neighbor was sick or in distress he never was in need long if anyone could help. We had faith and confidence in each other.

Zetta Farmer
Pratt, Kan.

Newsmongering Shoemaker

If my father were living he would be 116 years old. He used to tell about the pleasure his family got from the visits of the traveling shoemaker every winter. The whole family looked forward to his visit and they had the hides tanned and ready for him. He made shoes for every member of the family and often stayed several weeks.

They all were proud of their new shoes, but the greatest pleasure was derived from the news and gossip about all the families for whom the shoemaker had been working.

Mrs. M. O. Parrish
Vandalia, Mo.

Walking West

My parents never turned away anyone who wanted refuge for the night. Many is the night we had so many travelers with us that beds were piled down all over the floors.

One family in particular stands out in my memory. A man, his wife and three children were walking thru to western Kansas. Their team had died on the way, so they had bundled up some clothing, bedding and a few necessities and were walking thru to relatives who had homesteaded out west.

The man carried their possessions wrapped in a bed sheet that to my childish eyes seemed such a huge bundle. I marveled that he was able to carry it. The wife carried the baby about nine months old, a frying pan and a coffee pot. A little girl of five and a boy of seven walked by her side. My parents often wondered if they reached their destination or if they might have been captured by Indians.

Myrtle C. Thomas
Oakwood, Okla

CHAPTER 11: Pioneers Had Funnybones

End of a 'Raider'

We children used to sit spellbound by the hour and listen to Grandfather tell of his experiences as a pioneer settler in the Platte Valley of Nebraska. He settled in the days when Indians still were making raids on settlers.

One night word came that the Indians were coming. The men gathered in one of the homes and sat quietly with their guns loaded waiting for the expected attack. Finally, they heard a noise and saw movement in the bushes. They opened fire. Then all was quiet for the rest of the night. The next morning they found an innocent-looking calf in the yard dead, and full of bullet holes!

Mrs. Orville Hunt
Box 134
Wauneta, Neb.

Jolt For Johnny

My folks lived in a log cabin in Nebraska in 1882. Some Indians were camped on a river about a quarter of a mile away. One of the men came to our cabin every morning and begged for food for his squaw and papoose.

One morning Father saw him coming and grabbed my brother and me and put us in bed. I began to bawl and yell. When the brave stepped in (he never knocked—just pulled the latch string and walked in), he went over to the stove to warm his hands. He said, "How."

"Not so good, Johnny," Father said. He pointed to me crying in bed and said, "Smallpox!" Johnny grabbed his nose, let out a whoop and ran for the door. He left it wide open and ran all the way to his camp.

That afternoon we watched the Indians moving up the road, but they detoured away around our house. They had a new wagon with red wheels, the first I'd ever seen. Their teepee poles were dragged by ponies and the squaws walked.

As Father fixed the latch on the door, he laughed heartily and said, "Well, that was worth it!"

C. W. Martin
Litchfield, Neb.

Settlers in a Hurry

One day a pioneer boy herding cattle in the Blue Hills of Kansas saw two dark-skinned women picking wild plums. He thought they were Indians and started an Indian scare among the settlers.

He warned a number of families and they gathered up a few belongings, loaded them in their wagons and started a mad race to the blockhouse on the north side of the Solomon. The blockhouse had been built by earlier settlers in case of Indian raid.

The road from Pittsburg was strewn with all manner of belongings—kitchen utensils, hats, sunbonnets, tubs and washboards—lost in the frantic rush to the blockhouse.

My parents had not been warned, but they heard the wagons go by all thru the night and wondered about it. The next day the settlers went back to their homes, much relieved and gathering up their lost possessions all along the way.

Myrtle Thomas
Oakwood, Okla.

Mama's Rolling Pin

My mother was a wife and mother at 18. She lived on the outskirts of a village on the Wisconsin side of the Mississippi. A tribe of Indians came from the north and camped on the other side of the river, and when the water froze, they made nuisances of themselves as they came begging for food. Food was not too plentiful among the settlers and they hated to part with it.

Father had put some turnips in the cellar but they had proved to be so strong that the folks decided that is what they would give the Indians when they came begging.

One morning Mother was baking bread when four Indians, three women and a man, just opened the door and walked in. They never knocked. Mother had a white pine floor in her kitchen and always kept it immaculate. The Indians tracked in mud and stood there with muddy water dripping onto Mother's clean floor.

She gave them their turnips, but still they stayed. Mother knew they smelled the bread baking. She left it in the oven as long as she could, but finally she had to take it out or it would burn. She thought she probably would have to give them one loaf. As she put the pans on the back of the stove after turning the fresh loaves out on the kitchen table, she saw, out of the corner of her eye, the squaws each take a loaf and put it under her blanket.

The man reached for the last loaf. That's when Mother exploded. She grabbed her rolling pin and cracked him sharply across the knuckles! He let out a howl and bounded out the door. When he caught up with the squaws, he gave one a cuff and took her loaf away. Mother watched them go down the road tearing big chunks out of her bread and eating it hot.

When Father came home, Mother told of her experience. Father was frightened and told Mother never to hit an Indian. After all, they were only half civilized and there was no telling what they might do for reprisal.

The day the Indians broke camp to move on Father said, "Wrap up the baby—I'm going to take you down to your mother's to stay. Don't leave until I come for you tonight." There were no reprisals, but my father worked in the general store and knew the ways of the Indians. He knew that Mother's act could have been dangerous instead of funny!

Leona Haskell McDaniel
433 Lincoln
Topeka, Kan.

Survival of the Swedes

We once had a Swedish neighbor whose parents had brought their family from Sweden when she was 11 years old. They had joined a Mormon caravan of covered wagons intending to settle in Utah. At Nebraska City disaster struck the caravan and typhoid took many lives, including that of Mother Hansen.

In the spring, the two older boys went on with the caravan, but 11-year-old Ellen, her father and the younger children remained to settle in Nebraska. They traveled north to Omaha to make their permanent home. They arrived on the outskirts of town one summer evening. Ellen and her father took turns keeping watch over camp that first night.

Just before dawn, ear-splitting yells and shots broke out in the town. The Hansens had been alert for an Indian attack, so they quickly abandoned camp and all ran to hide in the tall brush. All day shooting and shouting rang out, and the family crouched in hiding without food or water. Toward evening the noise lessened and at nightfall it died away.

Venturing back toward camp, Father Hansen met some of his new countrymen. "How many Indians did you get?" he inquired.

After much gesturing and explanations made difficult by the language difference, the Hansens learned that they had survived— not an Indian attack—but their first Fourth of July in the United States!

Mrs. Frank Wacha
704 Gold St.
Schuyler, Neb.

Grandma Got Her Way

When my great-great-grandmother was preparing to move from eastern Ohio to "away out west in Indiana" she longed to bring several items that her husband said they absolutely did not have room to bring. Among these things was a little fire shovel her husband's brother had made for them in his blacksmith shop, but my great-great-grandfather flatly said no. Nevertheless, when they reached their new home there was the little shovel and there was Great-great-grandmother sewing up a seam in the featherbed!

Mrs. Eyman Turner
R. 1
Portland, Ind.

Took the Bread to Bed

Jim and Lucy were homesteaders in western Nebraska. One winter Lucy developed inflammatory rheumatism and was forced to take to her bed for many weeks. She was painfully ill, but not dangerously so. Jim found it necessary to leave her alone for long hours every day as he augmented their slender income by riding herd for a neighboring rancher.

Lucy was unable to tend fires, but with plenty of bedding she kept comfortably warm while Jim was away. However, day after day Jim tried to make light bread, mixing it before he left for work, wrapping it well and planning to bake it when he came home. It didn't work. The little homestead shack was too cold, and the dough would not raise.

At last, when both were utterly weary of eating biscuits, Lucy hit on the idea of having the bowl of bread dough tucked in beside her. That turned the trick. The warmth of her body kept the dough warm—and they could have light bread again!

After this success, Lucy cast about for other ways of utilizing her illness. In early spring she asked Jim to save the eggs from their tiny flock of chickens. She took two dozen eggs into her bed, kept them close to her body, and turned and shifted them as carefully as ever a mother hen could do. Twenty-three fluffy yellow chicks were hatched. By that time Lucy was enough improved to get up and care for them.

Nelle Portrey Davis
Welcome Ranch
Bonners Ferry, Idaho

White Hair's Peppermint Joy

When Father and Mother were first married he worked in Uncle John's trading post where they sold about everything—traps, skates, sugar, kerosene, piece goods. The Winnebago Indians used to come down to our village and camp across the river and trap all winter.

There was one old Indian who lived on the fringe of the camp by himself and to whom the other Indians paid no attention. One day Uncle John asked the chief why the old man seemed to be excluded from the group. The chief touched his forehead, meaning that the old man, White Hair, was simple and that they had some superstition about it.

White Hair was very fond of Uncle John and always sold his furs to him. He had one trait that seemed to me to be pretty smart. Every time he sold his furs he left two or three for credit. "Next year maybe not catch 'em," he'd say.

Uncle John always gave White Hair a little bag of peppermints when he brought in his furs, and this pleased the old man very much. Peppermints in those days were about twice the size of our mints today and were very potent.

White Hair did not come back the next winter and not until the following year did he appear. He explained that he had been "long time sick." He was glad to claim the credit he had left in the years before and to get his peppermints.

He kept saying, "Candy make well," and Uncle John couldn't understand what he meant. Then he remembered that he had given White Hair peppermints the last time, just before he was sick. He gave the old man an extra big bag of the mints, and White

Hair went away beaming and saying, "Present make me happy inside," which became a by-word in our family for a long time.

Leona Haskell McDaniel
Topeka, Kan.

Her Kitchen Prestige (Reprinted from Capper's Weekly)

My sister married a homesteader who had been batching and knew how to make baking powder biscuits. She always had made biscuits with soda. He claimed that his biscuits were much better than hers and offered to prove it by baking a batch.

The bride couldn't bear the thought of having her husband outshine her in her own province, so she emptied the baking powder can and filled it with flour! Of course, the biscuits were an utter failure. The husband couldn't understand it, but conceded that his wife must be a better cook than he. My sister hid one of those "rocks" and kept it all of her life, until she was 84. Now I'm 84 and I still have that biscuit!

Mrs. L. B. Boies
Astoria, Ore.

Grandfather's Sin (Reprinted from Capper's Weekly)

Grandfather brought Grandmother to Kansas when she was a bride in 1868, and they set up housekeeping in a two-room log house. The tale I remember best is a sticky story that almost parted Grandpa and Grandma—for a time.

Grandfather always raised sorghum and made a barrel of molasses for the winter. Every pioneer family had a barrel of sorghum in a corner in the kitchen. One Sunday morning Grandfather didn't feel well, so he decided he wouldn't go to church. The sorghum pitcher was empty, so Grandmother put it on the dirt floor and turned on the spigot of the sorghum barrel. It was winter and the sorghum ran out so slowly she couldn't wait for the pitcher to fill. She told her husband to watch it and turn off the spigot. Then she took the children and drove the team to church.

Grandfather sat down to read and soon fell asleep. The pounding of the horses' hooves as Grandma returned from church woke him. With a start he remembered the sorghum. He rushed to the kitchen door just as Grandma was opening the outside door. Both stood looking down at sorghum inches deep all around the barrel. Grandfather spent that Sunday afternoon shoveling out dirt and their winter's supply of sorghum.

"And," I remember hearing Grandfather say, "a dirt floor is mighty hard to shovel up after it has been packed for years. Then I had to carry in clean dirt and pack it down. Emily was awful hard to please that afternoon!" And he winked at Grandma.

"John swore he'd never miss church again," Grandma said. "And he never did."

The neighbors heard about the loss of their sorghum and many of them came to call bringing buckets of the sweetening from their own barrels. I live on the farm where this incident happened.

Mrs. Virgil Sellers
Weir, Kan.

Carry and the Choir

My grandmother came to Kansas from Pennsylvania in the early 1860s. She insisted that her family bring her cherished organ with them to a farm near what was later known as Vanango.

Grandmother became the organist for the Sunday meetings, and during the week the neighbors gathered at her home to practice the Sunday hymns. One girl, whose father was "well-fixed," had an alto voice and loved to take a leading part, altho she was terribly "falsetto," my mother said. Often she got away from the alto notes and her voice went off key into a squeaky soprano range. Grandmother often commented that the effect was mouse-like.

One evening when the choir had gone home, Grandmother said in an exasperated tone, "Carry sounded more like a mouse than ever!" She sat down at the organ to mimic Carry. She had no more than played the opening notes of the alto part when an awful squeaking began!

Yes! A mouse was caught in the reeds of the organ and the shrill effect had not been all Carry's fault. For years this was a source of mirth to these pioneer women.

Anne Pafford
Salina, Kan.

Pioneer Holiday

Away back in 1856 my father was a single man batching with a friend on a homestead in Nebraska. The two men attended Sunday School in Table Rock, ten miles away. One Fourth of July, a Sunday School convention was scheduled in Table Rock and a prize was offered for the most interesting delegation.

Father had a yoke of oxen broken to ride—only trouble was that one of the team was not an ox, strictly speaking. Father's bachelor friend dared him to ride old Mooly and head their delegation, and he did just that. He rode the animal that had been all curried and beribboned. A lane was roped off before the judges' stand, and Father said that there was yelling and men threw their hats into the air as he rode by. He was afraid Mooly would bolt, but he just put his head down and stoically endured the holiday mood of a pioneer Sunday School convention. And Father's delegation took the blue ribbon!

The only trouble was that Father's sweetie was angry and gave him, as they said in those days, a good raking.

Charles Smith, Jr.
Nora, Neb.

Grandpa and the Papoose

Samuel Stark and his wife, Elizabeth, and their small daughter came to what is now Kansas City in 1855. My father, Eli, was born there in 1856. It was called Westport Landing and Grandpa Stark was a partner in a small grocery business there.

Indians flocked to Westport Landing by hundreds. In summer they would strip off their buckskin and plunge into the muddy Missouri, Grandma said. Squaws would toss young children into the shallower water so they could learn to swim. The mothers apparently felt no fear that the children would drown. Grandmother said the bucks' wet bodies gleamed like polished bronze when they emerged from the river. She remembered especially the squaws' great love of red calico.

Westport Landing was a wild, unkept place then and fences were unknown. Cattle and hogs roamed and foraged at will. Grain was scarce and the animals often were ravenously hungry.

One day a squaw came to Grandfather's store and removed the papoose from her back and set it down outside the store. A hungry sow came along, sniffed the skin covering around the baby and decided it would make a good dinner. She rooted it over and pitched in with tusks and hooves, luckily beginning at the baby's feet.

The infant screamed and Grandfather Stark ran out and with heavy boots and a club fought off the hungry animal and rescued the papoose. Indians appeared like magic and when they understood what had happened, the mother insisted upon giving the baby to Grandfather. It was with difficulty he convinced her he could not accept the gift.

To show their gratitude, however, the chief had a sign printed over my grandparents' home and whenever Indians came to Westport Landing they brought gifts of fresh meat for as long as Grandpa lived there.

Charlotte Stark Offen
El Dorado, Kan.

Wedding Pranks

My father was an orphan boy living in the home of my grandfather. My mother was the 16-year-old daughter of the family. The two did their courting around the fireplace in the evenings with the whole family present.

One night Father wrote on a slate and asked Mother to marry him. She took the slate and wrote, "No." But the next night she wrote on the slate that she would take back what she had said the night before.

The next year they were married. After the wedding supper was over and bedtime came, the women put the bride to bed. Then the men came in and put the groom to bed with her. The young folks tormented them and stayed in the room all night.

They wouldn't let the couple get up and dress but neither would they leave the room themselves. They brought food on trays and kept them constant company, singing, dancing and playing the fiddle right there in the bedroom. The young guests took turns going to another room to sleep, but would not let the bride and groom alone. This was 85 years ago.

Vida Ward
Miller, Mo.

Bustle Banks

Back when the Ozark hills were first being settled, my grandfather was county collector for Newton County in southwest Missouri. There were no banks in this territory in those days, and the collector was responsible for the money he collected.

My grandmother solved the problem by sewing the greenbacks into the bustles she and her daughters wore! They never told where the money was kept until years later. But the funds were safe and ready for use whenever the county needed them!

Mrs. Earl Brown
Lanagan, Mo.

Horned Toad Hullabaloo

This little story of pioneer days was told to me by the grandmother in a home where I was working. She was in her 80s and spent most of the day in her favorite rocking chair where she dozed and dreamed of days gone by.

One day, sitting in her chair, she started chuckling. Then her chuckle turned into a hearty laugh. I said, "Grandma, tell me what is so funny so I can laugh, too." This is the story she told me:

Grandma and her husband came to Kansas in the early days. Her husband was a school teacher, so they hired a young man to break the land for them. In those early days, horned toads were very plentiful and ran everywhere in the fields. Men had to be careful because sometimes the toads, in their dash for freedom, would run up their trouser legs.

One day the young man was in the field busily hoeing away. All of a sudden, he felt something in his trouser leg, half way between his hip and his knee. A horned toad for sure! Quickly, he grabbed the lump on the outside with one hand, took his knife from his pocket with the other hand and hurriedly cut out the lump, trouser and all, and gave it a quick toss from him.

As it flew thru the air, he saw that it was not a horned toad at all, but his cherished plug of chewing tobacco! It had worked its way thru a hole in his pocket. He lost his tobacco and a big circle from his trouser leg—and years and years later, Grandma and I laughed until tears came to our eyes.

Mrs. Ernest R. Davis
Salina, Kan.

CHAPTER 12: When A Wagon Was Home

Enough in Her Time

I truly know what it was like to live in a covered wagon. When I was a girl I went with my parents on a leisurely trip from Nebraska to Arkansas. We lived in the wagon for a year then. And for eight months longer, until we got our dugout finished, we continued to live in the wagon and a tent.

As a bride I went with my husband in a covered wagon to the Panhandle of Oklahoma where we took up a homestead and lived in the wagon and a tent until our sod house was finished. I was so proud of that house and so happy when we moved in! My happiness didn't last long, tho, because the sod house caved in! Luckily, no one was inside. So it was back to living in the wagon and tent again.

I never expect to have to live in a covered wagon again, but I know how.

Mrs. Mary DeHaven
2600 S. Washington
Kennewick, Wash.

List of Supplies

A barrel of crackers, a barrel of sorghum, a barrel of flour, a keg of kraut and some cured meat were some of the things we loaded into our two wagons in 1888 when we moved from Ohio to

Arkansas where Father had bought a farm. We had three cows tied on behind and took our dog, Tige.

When we came thru Indiana we stopped at a farm and asked if we could camp for the night. The farmer said we could and Mother went about fixing supper. I was quite small and my older brothers thought it was funny for me to yell, "Hurray for Harrison!" He was a presidential candidate that year. I piped up with "Hurray for Harrison" in front of the farmer. The old fellow was very angry and ordered us to move on!

When we got to the place Father had bought, the rainy season had set in and the whole place was swampy. All of us took malaria and one season was all we spent there.

Mrs. Esta Hicks
Roselle, Mo.

Tornado Hit Wagon Train

Our family, with six other families, made up a wagon train moving from Nebraska to Colorado. There were no roads across western Nebraska and eastern Colorado at that time—just trails across the prairies and no bridges.

Each family had a spring wagon tied to the back of the covered wagon, and each family had a few cattle. My eldest sister, 14, and two brothers, 12 and 10, drove our cattle.

One noon when we were camped getting dinner, a tornado hit us. When the men saw the dark clouds coming, they turned the wagons with their backs to the coming storm. Our wagon was the last one to be turned. Mother and all of us children were in the wagon. Before it could be turned, it was blown over. My youngest brother got a few scratches, but the rest of us just scrambled out unhurt.

Another time a hailstorm drove the cattle back into a river we had just crossed. Still another time the cattle herders were lost with the cattle and went a whole day without food before they found us.

But we had good times, too. After the evening meal was

cleared away, we would sing songs and hymns and have evening devotions. On Sundays we camped all day. We studied the Sunday School lesson and had a worship service. The afternoon was spent visiting and doing little chores that had to be done.

Mrs. George Schlegel
Otis, Kan.

Overjet—Standard Equipment

In 1887 my parents were living on a rented farm in Iowa and decided to go to Nebraska and take a homestead. Late in the summer they sold most of their stock and farming implements but kept a young team, a few cows, two colts and with a new wagon and set of harness got ready for the trip.

There were nine children from 20 years old down to the baby, seven months old. Father built an overjet to fit over the top of the wagon box. It extended over the wagon box a little more than four feet. Slats were put across so they could put the bed tick and quilts on it. This made a good comfortable place for the girls and small boys to ride. Father and Mother rode on the spring seat.

Will and Henry, the two older boys, drove the cows. They could ride the three-year old horse and would take turns riding and walking the 300 miles.

My parents took just what they thought the most necessary with them—a cook stove, Father's tool chest, cooking utensils, vegetables, grain for the horses, and a wall tent.

I was only five years old when we made this trip and now I am 82.

J. B. Brubaker
Springfield, Mo.

Human Clothesline

It was a long, long trail a-winding that led us from Indiana to drouthy Kansas. There were nine of us, one a small baby—far too many to start so far in a covered wagon. It was a long, tiresome trip and it seemed to me it rained most of the time. We children slept under the wagon, and the rain would run under the wagon and get our beds wet.

With all the rain, there was no way to dry the "squares" for the baby. My sister and I had to be the clothesline and hold the diapers up before the campfire to dry them. We disliked this job very much. Some days we would make camp to get the washing done and let the horses rest.

I remember one blizzard after we got settled. We had plenty of fuel, but no water. We would open the door and dig out snow and melt it. One dry season the folks planted turnips and fall rains came in time for them to grow. We lived mostly on turnips that winter. I haven't cared for them since.

Mrs. F. Modlin
Burr Oak, Kan.

One Wagon West

The snow was a foot deep that morning when Mother, Father and we six children started on the long trip from Iowa to the "Golden West." No, not California, but Oklahoma where Father had relatives who already had taken claims.

Our earthly belongings were stored in the new wagon with double sideboards, plus an overjet that made it wide enough to allow a bedspring to be placed in its roomy interior.

Mother's new Singer sewing machine was carefully stored in the bottom of the wagon bed. There also was a huge green box with Mother's nice things, including yards and yards of material. I'll always remember the magic of the things that evolved from Mother's old green box during the dreadful years ahead. I saw my dear Mother struggle to keep things "nice" the way we had been accustomed to them at home.

The two-room sod house had a dirt floor and was a trial. Water was hauled in barrels from a spring seven miles away, so we saved every drop of the precious liquid.

But back to the wagon. We crossed the Missouri River twice on a ferry boat—a thrilling experience for us youngsters. During the trip Father had a chance to trade one of his team of beautiful black horses for two mules. He figured that two mules would be better than one horse. Mother cried when he traded because she thought the vicious things would run away with us. One of the mules turned out to be practically worthless.

One mule was tied on behind the wagon, and we older children decided to make good use of old Long Ears. We would take turns stealthily crawling thru the hole in the back where the wagon sheet was drawn together with rope. We'd grab old Long Ears and slide along his neck onto his back for a nice ride. We didn't have permission, but Father never seemed to notice. I still wonder if he didn't just enjoy our stunt.

About half way to Oklahoma we stopped over for three weeks at Grandfather's house. Mother had become ill, so we all rested.

I remember going across the Ozarks. The roads there were terrible. It seemed we rode for days to make a few miles because there were so many steep mountains and sharp curves. We children would take turns at seeing which one could walk the farthest. We would stop and break rocks into pieces because we had heard that there were golden nuggets in "them thar hills." Sometimes we would find beautiful rocks and want to keep them, but of course we had too heavy a load for such foolishness.

We arrived at our uncle's in northwest Oklahoma in early April. It was a cold, clear day, but I remember how beautiful was the morning of our arrival. When the sun came up, the world was like a fairyland. Rain had frozen on the tall, dead bluestem grass and everything was glittering and sparkling.

Mrs. A. L. Miskimon
338 1/2 N. Main
Wichita, Kan.

Lost Their "Roof"

I came to Oklahoma in a covered wagon with my parents, and later traveled to New Mexico in one when I was married.

We had an overjet with springs and mattress that made a nice bed. Bows and wagon sheet were fastened down good (so we thought) and we had a bachelor stove.

One night we had a terrific storm with wind and hail. The wind tore the wagon sheet loose, but my husband tied it down again. When we started out the next morning the wind was from the northwest and came right into our faces and into the front of the wagon. We had not gone far when that wet wagon sheet caught full of wind and away it went. This scared the mules and we had a run-away! The stove was jerked out of place, and when we finally got the team under control, we had to put out a fire in the wagon.

The river didn't look bad at all, but when we got about two-thirds of the way across it suddenly got deeper. The wagon bed began to float. We never could figure what kept it from floating downstream unless it was the hand of God. When we got nearer the edge the wheels settled in place again and the wagon was still hitched to the mules. We enjoyed our trip in spite of our difficulties.

Mrs. S. R. Moore
Harnmon, Okla.

Lou's Lost Love (Reprinted from Capper's Weekly)

My grandfather's parents came from Ohio to Kansas in a covered wagon. Their daughter, Lou, was engaged to John, a neighbor, who had left for St. Louis two weeks earlier with another wagon train.

Lou and John were to meet in St. Louis and be married. When Grandfather's party arrived, John was not at the pre-arranged meeting place and no one from his wagon train could be located.

Some of the family did not like John anyway and insisted that he had skipped out. They finally made Aunt Lou go on with them. During the trip another young man had been attentive to Lou, but she had given him no response. After she became convinced that John had jilted her, she was heartbroken. Six months later she married the other man.

It was two years afterward that John found Lou and discovered that she was married and the mother of a small son. He had been stricken with a terrible fever before he reached St. Louis and had hovered between life and death for weeks. He had been taken to a small village and there were no mail facilities thru which he could let anyone know about himself. When he recovered he joined another wagon train that traveled first to Omaha and then down into Kansas.

Both Lou and John were unhappy about their blighted romance, but in those days marriage was for better or for worse. Aunt Lou became the mother of six children and was known as a good wife.

Mrs. Albert Hay
Onaga, Kan.

Cussing the Tar Bucket

My grandfather had a half-brother who made wagons. There was very little iron about those old wagons, and what there was was beaten out on an anvil with a sledgehammer by hand.

The wagons were called tar grinders because, instead of putting grease on the spindle, they used tar. The early settlers carried their tar buckets swinging on the coupling pole. If they ran out of tar and the spindle ran dry, you could hear the wagon screeching a half mile away. This was called "cussing the tar bucket."

J. O. Bowman
Trion, Ga.

Breakfast on the Wagon Tongue

I wasn't quite four when we traveled by covered wagon from Kansas to a small mining town in Colorado, but I remember those chilly early spring mornings when we sat shivering on the wagon tongue eating the breakfast Mother had cooked over a campfire.

My sister and I would get so tired riding all day that Father tied two ropes to the back of the wagon and would let us out to walk if each of us would hold onto a rope. That helped us to keep up with the wagon and not get too far behind.

Not long after we got settled in a little one-room shack, all three of us children came down with scarlet fever. I've heard Mother tell how her back nearly broke holding one of her three sick children most of the time with only boxes to sit on. How she longed for a chair!

Ivy Schieck
Wheatland, Wyo.

Jennie's Alarm

Grandmother Jennie lived in a covered wagon box with her two little boys while Grandfather Willie went with some other settlers to haul wood into the Territory to trade with the Indians for food.

A neighbor had wanted their claim for a son not old enough to homestead for himself. He had threatened Grandma and told her he was going to scare her and the boys half to death some night so they would leave.

"Well, come right ahead, Mr. B," Grandma said. "I have a gun and I know how to shoot!"

Each night until Will returned, she put a row of pie pans and other noisy kitchenware around the wagon bed to warn her if anyone tried to get in.

Mrs. Robert Williams
Winfield, Kan.

CHAPTER 13: Dugouts, Soddies, and Shacks

Buffalo Grass Roof

Long, thick buffalo grass was the floor in my grandparents' sod house when they settled in Nebraska. This grass was the stomping and chewing ground for millions of fleas.

The sod house had a plain board roof covered with sod and clay, and two board doors with latch strings that were always out to friend and stranger alike.

One time when it rained and the roof leaked, my mother and her sister spent the night in the deep window seat, the only dry spot in the soddy. In the morning Grandmother fastened an oilcloth above the table to prevent the breakfast from being mixed with clay.

Mrs. Marvel B. Beery
107 W. Moneta Ave.
Bakersfield, Calif.

Heated With Prairie Hay

My first memories are of our little sod house in the sandhills of Nebraska. Two beds were set end to end along the north wall and the four-lid stove occupied the southwest corner. Apple crates were tiered to make shelves. The black oak drop-leaf table had its place between the stove and cupboard. The house was set close to the side of a sand hill and faced the creek on the east.

In the winter our fuel was prairie hay. A base was fitted at the front of the stove over the ash pit and a burner, oval-shaped like a wash boiler, about three feet high, was packed with hay and turned upside down and fitted in the rim of the base. A burner well packed with hay would burn two or more hours. Two burners made it possible to have another ready when one burned out.

Mother made our Christmas gifts. A match box covered with pretty paper and decorated with pictures from the seed catalog was one of my treasured gifts. Another was a cardboard star covered with the tinfoil from a tea package.

Father liked to make things for us, too. From the pieces from one of Mother's broken-down wash boards, he made very clever little wash boards for us girls. With an old cultivator he made my brother the nicest snow sled in the locality.

Vesta Millemon
Pond Creek, Okla.

Log Home

I am past 83 and nearing the end, but I well remember when we came to Aspen, Colo., to pioneer from Kewanee, Illinois. We came over the Independence Range by stage part way, part way on wheels and the rest of the way on a sled drawn by four spirited horses.

We upset once in a snowdrift. My youngest sister, Kate, was only seven weeks old. Just imagine my mother's plight carrying a seven-week-old baby over that route! I dare say few today are her equal. She passed away at age 78. God bless her!

My parents pre-empted 160 acres on Snowmass Creek. There were no roads, no bridges and no fences. Our cabin was built of round logs, and the lumber for the door, table and benches was hewn with a broad ax. In season, we had wild fruit—Oregon grapes, chokeberries, wild strawberries and raspberries. There was plenty of native trout, and deer and elk were available to those who could hunt. We couldn't, but we had good neighbors.

There were no schools near, so we moved to town for the winters. There were heavier snow slides in those days and many lost their lives.

James MacKenzie
Glenwood Springs, Colo.

No Dugout for Her!

Father went from Missouri to file on a place near Clinton, Okla., in the spring of 1892. Mother took my two brothers, my little sister and me to Arkansas to live with her sister while Father prepared a place for us to live. He broke ground and put in some crops.

Mother and we older children picked berries and earned enough money to buy a cow. Mother also got an old hen and raised some chickens. Father came for us before Christmas in a covered wagon. We had a heavy load and the roads were bad. It took five horses to pull our wagon. We children took turns driving the cow and two calves. Our cow had given birth to a calf and we had bought another one.

Sometimes we were so cold we almost froze. Mother took her beautiful, prized quilts to wrap us with. At night we had to sleep on the ground and Father would spread a wagon sheet over us. Sometimes we woke covered with snow.

When we came to the Canadian River, Father borrowed another wagon to help ford our things across. On the first load he took me, the dog and as many belongings as he could safely put on one load. He had to tie the wagon bed on to keep it from floating downstream. I was left on the opposite bank when he returned for the rest of the family.

When we landed at our place, a load of wood was all there was on it. Most of the homesteaders lived in dugouts, but my mother said she didn't want to go into the ground until she was dead! So Father got a tent and a small house from a neighbor.

Mother raised lots of chickens from that old hen and the little chickens she started with. Quite often during those early days, we could hear the howling of the wolves, and the beating of tom-toms. We were not sure then that the Indians were friendly.

Mother and Father had to travel to El Reno for supplies, and usually it took four days. We children stayed with neighbors. In the summer of 1893, we had a log-rolling and had a new log house. Father had plowed for the Indians in exchange for the logs.

Our schools were sod houses, and on cold days we would play blind-fold inside. It was plenty dusty! On our first Easter, everyone took dinner and spent the day at the old sod schoolhouse. It was crowded, but everyone was happy.

Rosa Riddle Jackson
1213 N.E. 20th St.
Oklahoma City, Okla.

Home of "Firsts"

My maternal grandparents joined the trek from the borderline of civilization in Wisconsin soon after the G.A.R.s were mustered out following the Civil War. They loaded their three boys and four girls into covered wagons drawn by ox teams Tom and Jerry, Dave and Darby, Duke and Dandy and Buck and Bright. It took 38 days of travel to reach Gibbon, Neb., where they spent the winter.

Their home on Cedar Creek boasted the first cook stove in the community and was the location of the first day school and first Sunday School. The first funeral of the new community was held there. It was for a young wife who had perished in the Easter blizzard of 1873 in their yard, so close to shelter. The roof of her dugout had caved in and she had gone seeking shelter. The first post office also was in my grandparents' home, and so it was the center of the community and news grapevined in and out.

Clothes were spread on the grass or hung on the plum bushes to dry. The children were given the task of watching them to see that they did not blow away, get chewed by cattle or that the grasshoppers did not eat them.

Mrs. N. D. Ickes
Page, Neb.

Cradle Lore (Reprinted from Capper's Weekly)

About 70 years ago when I was nine years old, I remember hearing pioneer women tell about ways they saved time on busy days.

One woman said she always put the big baby in one end of the cradle and the little baby in the other. As the big baby played, it kept jiggling the cradle and kept the little baby content.

Another neighbor said she put her baby in a high chair, put molasses on its fingers and gave it a feather to play with. That kept it happy for hours. A third woman said she put her baby in the cradle and gave it a piece of fat meat to suck. She tied a string to the meat and attached the other end to the baby's toe. That way if the baby choked on the meat, it would start kicking and pull the meat out.

Mrs. E. G. Caine
Indianola, Neb.

Father Made Furniture

My parents were married in 1876 and settled in midwestern Nebraska. They lived in a dugout near a small creek. They had a sod fireplace and a little stove for cooking. We burned cobs, cornstalks, sunflower stalks and chips from the cow pasture. When corn was eight cents a bushel, we burned some of that.

Father made a table and other furniture that was very sturdy, but not very polished. One time we had two kegs for Father and Mother to sit on at the table. A man came by who wanted a keg for pickles and he traded us a chair for one of the kegs. We children stood up at the table and didn't think anything about it because we were used to it.

Our parents didn't even have a table when they were first married. They sat on big pumpkins with the food spread out on a canvas on the bed. Once our lamp chimney got broken and Mother went to the smokehouse and got a saucer of lard. She buried a twisted cloth in the lard and lighted the end. We were perfectly contented with such makeshifts—at least, we children always were happy.

The first school my brother and I attended was in a neighbor's house. He had a two-room house and all the other homes had only one room. We had to furnish our own books and seats, so we had quite a variety. The teacher was a 17-year-old girl who was paid $15 a month.

One morning when we went to school, there was a new baby in the living room and water had to be heated on the topsy stove in the schoolroom. The teacher had us take our books and go out under the trees. We had no desks, but the teacher allowed us to sit at a table to learn to write. The next year our father and the neighbors built a sod schoolhouse on our place.

We were 75 miles from a doctor. Mother had a toothache for several months and there was no relief for it until spring. She had to wait until her baby was born and a month old, and then we made the long trip to the dentist.

One neighbor family had to burn hay for fuel. The father had TB and couldn't work He sat by the fire all day and twisted up bunches of hay to feed the fire. The children had no shoes except cloth moccasins their mother made.

Jessie P. Gentry
Stratton, Colo.

One-Room Dugout

In 1887 my parents took up a claim on the old Fort Wallace Reservation in Kansas. Our new home was a one-room dugout with three half windows, one door and a good Mother Earth floor.

Buffalo chips were our only fuel. Back of the stove sat a fuel box in which a good supply of the chips was kept along with a hatchet with which to break the large, solid chips.

If the supply became low, someone would take a gunny sack and go out on the prairie and pick up more. In the fall, my father would hitch a team of oxen to an old lumber wagon and spend several days on the prairie gathering chips for our winter fuel.

These were carefully stacked in the sod fuel shed.

How I loved fuel gathering days! I was allowed to ride in the wagon and sometimes I even got to drive the oxen. After there were no more buffalo chips we burned cow chips which were almost as good.

Mrs. Frances Ball
Hiawatha, Kan.

Hatchet Helped Get Homesite

When the Comanche Pool country opened for settlement, Father went to this new country. One day he was staking out a place near Coldwater where the half-dugout was to be located when a cowboy rode up. The cattlemen resented the encroachment of the farmers.

"What are you doing?" queried the cowboy.

"Planning to build a house," replied Father.

"You'd better move on," said the cowboy, cradling his gun in his arm. "You're not wanted here and it won't be healthy for you to stay."

Father was quietly winding up the twine with which he had been measuring for the foundation. He said, "You'd give a fellow time to gather up his things, wouldn't you?"

"Sure. Get busy."

Father picked up things until he was near the cowboy. Then with an uplifted hatchet in one hand, he reached for the man's gun with the other. The gun was released and the cowboy left without it. Father turned the gun over to the authorities and went ahead building without further interruption.

The country was wide, grassy plain, ideal for grazing, so it was no wonder there was strife between cattlemen and farmers. A herd law was passed requiring all stockmen to herd their cattle on open range and keep them away from crops and gardens.

The county surveyors were sent out to complete the land survey. Three young men lived in a tent near us while they were surveying. Mother furnished their meals. Sometimes in the evenings they would join the family for friendly conversation. Billie Thorn, the stoutest, used to quote these words: "When I'm a man, a man, I'll be a surveyor, if I can, and I know I can. I'll establish corners here and there, just to hear the people swear."

Vallie McKee
Anadarko, Okla.

Borrowed His Shack

This incident happened during the early homesteading days in eastern Colorado. Old-timers will remember that to hold a homestead one had to build a dwelling on the land and do certain other work. Most homesteaders stayed on the land only the length of time required by law and took a job, usually far removed, to earn a grub stake to finish the prescribed stay and to finance needed improvements.

Jim had built his little shack, and because he was a man of ambition and planned to make a real home on his land, he was not satisfied with a little tarpaper shack. Instead, he constructed a neat 12 by 16 house with clapboard siding and a shingle roof. He mounted it on runners so he could move it about his claim as he chose. Eventually it would be a farm work shop.

Then Jim went East to earn a few months' wages. He would come back in the spring and break a few acres of sod for a wheat crop. He got back to the prairies a few days earlier than he had expected, caught a ride with a neighbor to within a couple of miles of his claim and walked in after dark, carrying some groceries.

Tired from his trip, Jim took to his bunk at once. Some hours later he was awakened by the sensation of moving. At first he thought there was an earthquake, but finally decided his cabin actually was moving across the prairie! He could hear the clop-clop of the horses' feet and the low commands of the driver. A neighboring homesteader had decided to move the shack to his own claim during the owner's absence. He planned to prove his claim by having a dwelling on the land and then return the cabin before Jim came home.

The driver was one surprised man when the cabin door opened and Jim confronted him with a drawn .45. The would-be mover had no choice but to turn the team around and pull the cabin back. The experience did not sever the friendship of the two men. They lived neighbors for years, their children went to the same school, and Jim's eldest son married the neighbor's pretty daughter.

Nelle Portrey Davis
Welcome Ranch
Bonners Ferry, Idaho

Prairie Coal

In our early days on the prairies of Nebraska I can't remember that we burned anything for fuel except buffalo and cow chips. Sometimes we drove five or six miles to get a wagon load for winter use.

One young homesteader told his fiancee that he burned "prairie coal" when she wrote to inquire what he used for fuel. He was very vague as to what it was like. On their ride out to his homestead after they were married, she kept asking where the prairie coal was. Finally, he pointed out some to her. She was a dainty Eastern bride, and she was ready to go back East! Incidentally, she was distantly related to Queen Victoria. She carried on many years in her pioneer home.

Mrs. C. M. Clements
Highland, Kan.

Warm in Winter, Cool in Summer

When we came to our homestead in Barton County, Kansas, 76 years ago there was a little one-room block house on our place with a roof of brush and hay. Father had to get a neighbor to make the long trip to town to get some lumber for a roof.

Father built another room on the house out of sod. Always that room was nice and warm in winter and cool and pleasant in summer. Every stick of furniture we had was made by my father. He made a bedstead, a table, a chair for Mother and a bench for us children.

Later there was an extra little room where we kept corn for the hogs and chickens. My brothers had to sleep on top of the corn! We lived this way for ten years before we could afford to build a stone house.

Father had a team of oxen and Mother had to go along when he plowed to lead them. In winter it took one person's full time to cut the cornstalks and sunflower stalks to feed the fire. My mother refused to use prairie coal (cow chips).

We children had to walk two and one-half miles to school and our lunch usually was a piece of bread and some fat bacon. But we all grew up to be strong and work hard and every single one of us reached the 70s.

Louisa Wondra
Odin, Kan.

Life in a Soddy

My folks were married in 1884 and went directly to Miller, S.D., where they took up land and built a one-room soddy. My sister and I were born with just a midwife to bring us into the world. Our sod house had a floor, but many didn't.

We were 24 miles from town and Father had to stay over night each time and leave my mother alone in the soddy. There was a big Sioux reservation near and we saw many Indians, but there never was any trouble.

My father was reared in Philadelphia ant was not a farmer. Breaking the prairie sod with oxen and old blind horses (the only kind he could afford) was an agonizing experience. However, the soil was very good.

I was less than a year old when the blizzard of 1888 came and I was kept in bed all day. The bed was shared by my parents except when they had to get up to build up the fire or get something to eat. The vicious storm forced its powdered snow thru every possible crack. Father tried to reach the stable to feed his stock, but the building was obliterated by the white shroud of snow. When he gave up and turned to go back he could not find the soddy. It, too, was buried in a drift. Only by the grace of God did he find shelter again.

Some of the neighbors were burned out completely by prairie fires. The summer before my sister was born Mother was home alone except for me. She saw the fire coming and laboriously carried everything she could down into our little cave. As the blaze swept closer the wind changed suddenly and that saved the day.

After seven years of heart-breaking hardship, we escaped by prairie schooner. We traded our entire interest in the claim for one horse! Father fitted out another covered wagon and put Mother's high-backed rocking chair up front. She sat there and held the baby all the way back to Missouri!

Mabel M. Sturgis
Boise, Idaho

Dug Home in a Bank

I came to western Kansas in the fall of 1879 with my widowed mother, two sisters and three brothers. We had two covered wagons, four cows and half a dozen hens. The boys made a place to live in by digging a room about 30 by 40 in a steep bank. We fixed it up, and we really thought it was pretty nice!

Mother, two brothers and one sister took homesteads of 160 acres apiece. We had to haul our water five miles. Our meat for the first few years was mostly rabbit and antelope. Our family is all gone but one—and I will soon be 90!

Esther Andrews
Norcatur, Kan.

CHAPTER 14: Vittles and Duds

Prayed for Food and Water

I've heard Granddad tell many times about their wagon train running out of water ant nearly out of food on the way from Illinois to Colorado. One evening the women in the party insisted on having a prayer meeting to ask for rain. The men in the group thought it would be more profitable to send scouts out to look for water. Men on horses went in several directions.

Presently a rain started and all available pots and tubs were set out to catch the water. Still very little was caught. Then the women spread all the clothing out to catch moisture and wrung the water out into containers and spread the clothing again. There was thunder and lightning.

A flock of wild ducks flew over and some of them were struck by lightning. Five ducks dropped close enough for the women and children to get them. The men found no water or food, but when they returned to camp, both were waiting for them. Granddad said no one in the party belittled prayer after that.

Mrs. A. L. Rader
Loma, Colo.

Baby Had Milk

My grandparents bought milk from farmers along the way as they traveled by covered wagon from Nebraska to Oklahoma. One evening when they had camped for the night, Grandfather sent the girls to the nearest farm to buy milk. For the first time, they were refused. By this time the baby was so hungry he was crying.

Grandfather handed a pail to one of the boys and told him to go and milk a cow in a nearby pasture. Before long, the boy was back with milk for the baby.

Mrs. Howard Houston
Route 1
Boone, Iowa

Sugar Was for Company

My grandparents had to freight everything they did not raise on their homestead from Kansas City, 168 miles away. The trip was made about every six months, and Grandmother used to tell what a pleasure it was to get a few yards of material for a new dress.

Each member of the family was given one tablespoon of molasses at each meal. Sugar was so precious that it was kept in a sugar bowl tucked away in a dresser drawer. When company came, it was brought out proudly and placed on the table.

Gladys Norris
Hot Springs, Ark.

Thank the Lord for Rabbits!

My grandfather and his two eldest sons volunteered for service during the Civil War, leaving my grandmother and nine children on the homestead in Iowa. Mother said that she could remember that they lived mostly on cornbread and syrup, with rabbits for meat. The older boys at home, whose job it was to bring in the rabbits, used to chant:

"Rabbits tender, rabbits tough. God be praised, we've rabbits enough!"

Mrs. Myrtle Morphew
Roann, Ind.

Wild Greens and Rabbits

We lived mostly on wild greens and rabbits the first year or two we were in the Indian Territory. Sometimes our bread was made from bran we fed the horses. We got our drinking water from the little streams, and many settlers suffered from chills and malaria. This was 16 miles south of what is now Oklahoma City.

Ada Rhudy
Moline, Kan.

Hunger Fever

One year my grandparents, pioneering in Kansas, raised only one gallon of small potatoes. These were kept to make soup. I don't know what they ate the rest of the time. The government issued rations during the grasshopper year, but our family was too proud to get its share. The little they had went mostly to the children.

The women nearly starved, and my aunt got what was called "hunger fever." My mother also suffered the effects of that disease all her life. Later Mother had to pull water out of a well for 40 thirsty cattle. Women would not last long these days with such hard work.

Mrs. M. E. Rice
Emmett, Kan.

Wedding Feast

My father told of a pioneer wedding he attended when he was a young lad. The menu consisted of cornbread, molasses, sheep sorrel pie and creamed buffalo peas. There may have been other items such as rabbit pot roast and cornmeal coffee.

Father told me how they made cornmeal coffee. Molasses and cornmeal were put into a skillet and burned until the mixture was crisp. Then it was crumbled and brewed into a mixture called coffee.

Mrs. Arlo Howell
Fairbury, Neb.

Popular Family

The scarcity of good food, the utter loneliness and monotony of the prairie made life especially hard for the women. It isn't easy to be cheerful when the stomach yearns for "civilized" food. There was no milk or butter because the cows died of Texas fever. No fruit except wild plums and grapes. Very little pork. (Did you ever try to fatten hogs on prairie grass?) Most meat was wild game, of which one soon gets tired.

It was no wonder that when a new family moved in from Indiana bringing with them a load of good food, the word got around quickly. "Yes, the Sculls have pork and sausage, butter and lard, honey and everything just like back home!"

The Sculls had many callers to welcome them, and they were generous enough to send their new neighbors home with stomachs full of "civilized" food.

Mrs. Pruda B. Utley
407 S. Sixth St.
Arkansas City, Kan.

Sewed All Night

In the days of log houses, the spinning wheel and the loom, Grandfather had gone on a long trek across the plains to the gold fields of California. Grandmother was left alone to care for their little family.

In the long winter of 1855, the snow lay deep on the prairie. For many weeks the sun shone brightly around the little cabin without a trace of melting snow. It was bitterly cold and Dale, the eldest child, badly needed a new coat.

Grandmother rose very early and began her weaving. By night she had woven enough cloth to make Dale a coat. When the evening meal was over, Grandmother cut out the cloth and began her sewing. Elizabeth, the eldest daughter, sat by her and threaded her needles. The work was done by candlelight, and as the hours passed the candles burned out.

Undismayed, Grandmother continued her sewing by the flickering flames of the logs in the fireplace. Day was breaking in the east when the coat was finished. Grandmother was very, very tired—but proud and happy in the assurance that her eldest son and little helper had a warm coat.

The daughter, Elizabeth, was my mother.

Mrs. Emma Brawner
248 Eest Linden Ave.
Fremont, Neb.

Warm Petticoats

I was born in a dugout, but later Father built a stone house over the dugout. One winter my mother told my sister and me that she was afraid we'd have to stop school because we did not have enough clothes to keep us warm. But when we came home from school the next night, Mother had solved the problem. She had made us petticoats out of an old quilt she decided she could spare. We were proud of our quilted petticoats and showed the other girls. They were very, very warm. On bitter cold mornings, we donned our petticoats, Father wrapped our feet in gunny sacks and we trudged off to school.

Mrs. Will Sammons
Stockton, Kan.

Snacked on Jerky

My parents pioneered away back in a mountainous place in Oregon on the edge of a lake. All of our supplies had to be packed in on horseback. My father went out twice a year to get such supplies as flour, sugar, oatmeal, baking soda and salt.

We had plenty of fish and my father killed deer. We usually had a 50-pound flour sack or two of jerky (dried deer meat). It was delicious. We children would just get a hunk and chew it off! Mother's refrigerator was a place dug back under the creek bank like a shelf where the water ran about an inch deep. Once a rain came in the night and washed all the crocks of butter, cream and milk into the lake.

My father cut bee trees and we had wild honey all the time.

Mrs. Charles Pidgeon
Salem, Iowa

Indian Delicacy

My 72-year-old mother tells the story of the old days that is true, but hard to believe.

When she was a small girl, her family lived near some friendly Indians. The children visited back and forth. One day Mother was playing with the Indian children when their mother brought a live land turtle into the house. She put it in the oven and built the fire.

Mother said she could hear the turtle moving around for a long time, and then it was still. The women kept the fire going, and in the evening the Indians just scooped the meat out of the shell and ate it—all of it.

Mrs. Alfred M. Foster
216 Conark Courts
Conway, Ark.

You'd Be Ragged, Too! (Reprinted from Capper's Weekly)

My grandmother and her four little children lived alone in a dugout on their homestead while Grandpa worked in the coal mines in a different state.

One day Grandma saw a man with a blanket over his head coming toward their dugout. She was sure he was an Indian and was afraid. She could speak no English, so she told the oldest boy to stand by the door and hand the Indian a loaf of rye bread when he came to the door. She prayed as she gathered the infant in her arms and pulled the younger ones near her and braced herself to fight to protect them.

What a shock when the man opened the door and asked in Grandma's own tongue, "What are you doing with this bread, little boy?" The children fell on the stranger and hugged him. He turned out to be a neighbor from the nearby hills out looking for his sheep. He wore an old skirt over his head to protect his eyes from the midsummer heat.

When Grandma inquired about his strange garb, he replied, "Wait until you've been here a few years, you'll be ragged, too!" She was; the family wore out every stitch of clothes she had brought in the big trunk and when she lost her needle, she had to write to her mother in Europe to send her one. That day was a good one for Grandma because it was the day she met a true neighbor and friend.

Mrs. Anne Wesely
Cedar Bluffs, Neb.

Grandpa Was a Rover (Reprinted from Capper's Weekly)

My grandfather had an itching foot, a good trade, a moving wagon and a team of oxen. He was a cooper or barrel maker and he could get work in any community. He yearned to see all of the new places people talked about, and he did see them by slow stages.

He would load up his precious strawberry plants and his family and move to whatever new part of the country someone recommended. Grandmother went with him thru Ohio and into southern Missouri, then up to the southern border of that new territory called Iowa. There she balked. If he went again, he must go alone. Alone he went!

Grandmother was left in a strange new country with three children. The eldest was my mother, then 14. They had a house and a cow and very little else. What hard times they went thru before my mother got old enough to teach school and help out at home!

Mother had two linsey-woolsey dresses. At that time people wore red flannel underwear and because it would shrink so badly, the garments were aired on the clothesline oftener than they were washed.

My mother saw someone crocheting and yearned to crochet herself. She learned how, but it was over a year before she could afford to buy a crochet hook. She did shadow crocheting so she wouldn't forget how.

Mrs. Florence Gunn
Route 1
Hume, Mo.

CHAPTER 15: Grandma's "Receipts"

Johnny Cake (Reprinted from Capper's Weekly)

Once a band of Indians walked in on Grandma and a tall, stately leader ordered her to prepare a meal for them. Grandma did not want to lose her scalp with its lovely black hair, so she cooked the best she had. After they had eaten, the tall Indian touched her on the shoulder and said, "Good cook."

One of Grandma's recipes was for Johnny Cake.

"Two cups Indian, one cup wheat,
One cup good eggs that you can eat,
One-half cup molasses too,
One big spoon sugar added thereto,
Salt and soda, each a small spoon.
Mix up quickly and bake it soon."

Mrs. Robert Brenner
Irving, Kan.

Note: This recipe in rhyme was sent in by several women exactly as it is given above, so it must have been in circulation at one time. An improved version cuts down on the number of eggs and adds butter.

Yeast from Hops (Reprinted from Capper's Weekly)

Getting and keeping yeast was vital to pioneer homemaking. Here is a recipe for yeast made with hops.

"Add one quart of water to one pint of hops. Simmer until the strength is out, about 20 minutes. Then use the liquid to make a thick cornmeal mush. When it is cool, work in more cornmeal and then pat into little cakes. Dry the little cakes on plates and store them."

Mrs. Bessie Steeley
Iola, Kan.

Roast Goose in 1870 (Reprinted from Capper's Weekly)

This is a recipe for roast goose the way it was done in 1870.

"On the day before Christmas, kill a fat goose and dress it. Wash it well in a dishpan of hot soapy water. Rinse in a milk pail of cold water. Dry it thoroly and hang it up in the woodshed over night. Next morning early, mash a kettle of potatoes with cream and butter and a cup of chopped onion and lots of salt and pepper. Stuff the potatoes into the goose and sew it shut. Rub the skin over with salt and pepper and sage and put it in a not too hot oven. Dip the grease up every hour or so and save for cold-on-the-lungs and shoes."

Miss Ruth Rose
Lincoln, Kan.

Batch of Bread (Reprinted from Capper's Weekly)

This recipe for a huge batch of bread was taken from a cook book published in 1832. Seems to us that the woman who followed this recipe needed a whole day with nothing else to do and a pair of good strong biceps.

"Put a bushel of flour in a trough or a large pan; with your fist make a deep hole in the center thereof; put a pint of good fresh yeast into this hollow; add thereto two quarts of warm water and work in with these as much of the flour as will serve to make a soft smooth kind of batter. Strew this over with just enough flour to hide it. Then cover up the trough with its lid or with blanket to keep all warm.

"When the leaven has risen sufficiently to cause the flour to crack all over its surface, throw in a handful of salt, work all together; add just enough luke-warm soft water to enable you to work the whole into a firm, compact dough, and after having kneaded this with your fists until it becomes stiff and comparatively tough, shake a little flour over it and again cover it with a blanket to keep it warm in order to assist its fermentation.

"If properly managed, the fermentation will be accomplished in rather less than half an hour. Meanwhile that the bread is being thus prepared, you will have heated your oven to a satisfactory degree of heat with a sufficient quantity of dry, small wood fagots; and when all the wood is burnt, sweep out the oven clean from all ashes. Divide your dough into four-pound loaves, knead them into round shapes, making a hole at the top with your thumb, and immediately put them out of hand into the oven to bake, closing the oven door upon them. In about two hours' time they will be thoroughly baked, and are then to be taken out of the oven and allowed to become quite cold before they are put away in the cupboard."

Pearl Speck
Vici, Okla.

Corncob Sirup (Reprinted from Capper's Weekly)

Early Norwegian settlers living in log cabins in Wisconsin used this recipe for an imitation maple sirup. It tasted just grand on a stack of pancakes on a real cold winter morning—and still does. We know because we still make it when we are able to get the corncobs.

"Boil one dozen clean corncobs (red ones are best) from one to two hours in enough water to leave one pint of liquid when done. Then strain and add two pounds brown sugar. Boil as long as you wish as some prefer a thick sirup and others prefer a thin sirup."

Mrs. H. O. Daley
Parker, Colo.

Sheep Sorrel Pie (Reprinted from Capper's Weekly)

During pioneer times lemons usually were very scarce. To satisfy their taste for lemon pie, women learned to use the pink-flowered sheep sorrel that grew wild on the prairie as a substitute for lemons.

They used a regular lemon pie recipe but substituted a cup of shredded sheep sorrel for the lemons. The pie had a tart, luscious flavor similar to real lemon pie.

Mrs. Francis M. Wise
Hiawatha, Kan.

Acorn Bread (Reprinted from Capper's Weekly)

Early pioneer women learned from Indian squaws to make bread from acorns. Indians used to sneak over the Missouri River bluffs and gather bags of acorns to add to their food supply. The acorns were processed by long cooking in wood ashes to remove the bitter taste. Here is a pioneer recipe for making Acorn Bread:

"Shell acorns and cover with water to which one-half teaspoon of soda has been added to each quart. Boil until soft and wash through three or four waters. This removes the bitter taste. Spread on plates. When dry, grind in a food chopper. Take one cup each of barley flour and cornmeal, three cups acorn flour, one-half cup sugar, one teaspoon salt, two cups buttermilk, one teaspoon butter and one tablespoon melted fat. Mix into a loaf. Bake one hour in a moderate oven. Dark but delicious."

Mrs. Francis M. Wise
Hiawatha, Kan.

Vinegar Pie (Reprinted from Capper's Weekly)

My husband's parents came to the desolate new Wyoming country when their children were small. For many months they did not taste pie. There was nothing to make pie of—no fruit except buffalo berries and they were 10 miles away on the river.

They went to an auction sale and the woman gave Mother a

piece of the pie she had made for the auctioneer and bookkeeper. Mother was astonished to see and taste pie again.

"My goodness, Mrs. Hawkey, where did you get anything to make a pie? What is it?" That's how our family got this recipe for vinegar pie.

"Boil one-fourth cup vinegar and two cups of water together. Mix one cup sugar, a pinch of salt, four tablespoons flour and two beaten egg yolks together. Then add water and vinegar slowly. Cook until thick. Add one teaspoon lemon extract and one teaspoon butter. Pour into a baked crust. Top with meringue made from two egg whites and brown in a slow oven."

Melba Davis
Greybull, Wyo.

By-Guess-By-Gosh Gingerbread (Reprinted from Capper's Weekly)

Pioneer recipes were seldom as specific as recipes today, but this one for gingerbread from my grandmother's day really has me puzzled. She wrote:

"I always take some flour, just enough for the cake I want to make. I mix it up with some buttermilk if I happen to have any of it, just enough for the flour. Then I take some ginger; some like more, some like less. I put in a little salt and pearl ash, and then I tell one of my children to pour in molasses till I tell him to stop. Then the children bring in wood to build up a good fire and we have gingerbread for company."

No doubt Grandma got better results from this recipe than I would!

Mrs. E. A. Stowell
Boscobel, Wis.

Nellie's Suet Pudding (Reprinted from Capper's Weekly)

I have an old recipe book dating back to about 1850. It includes general information as well as recipes—such as a method for removing head lice! I was amused by a formula for a face lotion containing an ample amount of whisky. Imagine going to a party after using that!

Here and there in the book are recipes contributed by Nellie. I've often wondered who Nellie was. Here is Nellie's Pudding:

"Half a pound of flour, half a pound of treacle, half a pound of suet, the rind and juice of one lemon, a few strips of candied lemon peel, three tablespoons cream, two eggs. Chop suet fine; mix it with flour, treacle, lemon peel minced and candied lemon peel; add the cream, lemon juice and two well-beaten eggs. Beat the pudding well, put in a buttered basin, tie it down with a cloth and boil from three and a half to four hours."

In the book is a recipe for a ribbon cake that has left me puzzled. It is baked in three layers, a dark one for the center. The cake is put together with jelly while warm, and I quote: "Lay a paper over all and then a thin sheet on which put two irons. The cakes will be pressed in about two hours."

Mrs. Kenneth L. Chase
Burr, Neb.

Mistress of Her Oven (Reprinted from Capper's Weekly)

It took a bit of skill to be a good cook in the days before automatic oven controls. These instructions on "How To Heat an Oven" are taken word for word from a cook book printed in 1805. I've even left in the capitals in strange places.

"Some people consider it economical to heat Ovens with fagots, brush and light stuff. Hard wood heats it quicker and better. Take four foot wood split fine and pile it criss cross so as to nearly fill the oven and keep putting in. A Roaring fire for an hour or more is usually enough.

"The top and sides will at first be covered with black soot. See that it is all burned off. Rake the coals over the bottom of the oven and let them lie a minute. Then sweep it out clean. If you can hold your hand inside while you count Forty it is about right for flour bread. To count twenty is right for Rye and Indian.

"If it is too hot, wet an old broom two or three times and turn it round near the top of the oven till it dries; this prevents pies and cakes from scorching on top. When you go into a new house, heat your oven two or three times to get it seasoned before you use it.

"Bake the Brown bread first, then the flour bread and pies, then cake or puddings and last Custards. After everything else is out, put in a pan of apples. Next morning they will be deliciously baked. A pot of Beans can be baking back side, out of the way with the Rest."

Mrs. Dorothy Hofbauer
Ravenna, Neb.

Peach Leather (Reprinted from Capper's Weekly)

Peach Leather was a dried peach confection used long before canning was in common practice. I'm told that the method was brought from Africa by early slaves and Peach Leather was made by Negro mammies in the South before it spread to other parts of the country.

"In order to obtain the best results very soft peaches are used for Peach Leather, ripe to perfection, but not over-ripe. Wipe peaches off, seed and mash them to a smooth pulp, spread thin on platters and set in the sun. This should be well protected by fine mosquito bar from flies.

"When one side is dried to a slight brown, turn it with a knife so the other side will dry. Two or three days during the sunniest hours will complete the drying. The leather is then sprinkled with brown sugar, rolled up, wrapped in paper and placed in stone jars and covered closely. For serving, it is sliced through the roll."

Mrs. Lorin Best
Mission, Texas

Wild Grape Dumplings (Reprinted from Capper's Weekly)

I got this recipe from a pioneer neighbor, who got it from an Indian woman.

"Take two cups of wild grapes after stems have been removed.

Cover with water and boil about 15 minutes or until juice has been extracted. Remove from fire and strain. Add one cup of sugar and enough water to make one quart of juice. Return to fire and bring to a brisk boil. Then add the dumplings.

"To make dumplings, sieve together two cups flour, one-half cup sugar, four tablespoons baking powder and one teaspoon salt. Cut in four tablespoons shortening. Mix with three-fourths cup sweet milk to make a soft dough. Drop from a teaspoon, one at a time, chunks of dough about the size of a small egg, into the boiling juice. Cover and simmer for 15 minutes."

Mrs. Pearl Short
Altoona, Kan.

Prospector's Sour Dough

This recipe was used in the days of gold prospecting and by many hunters and cowboys far from home.

"Place one pint of flour in a pail with a lid. Add two tablespoons sugar and one tablespoon salt. Mix well. Stir in one and one-half cups water and beat to smooth dough. Add one tablespoon vinegar to the batter. Set pail in warm place where it will sour thoroly.

"When it sends up a sour odor, it is ready to use. Take nearly all of this and add flour, a little sugar and a teaspoon soda. Make a stiff dough. Put in your pan and let rise. Bake till brown. By varying the amount of flour used, you can make a stiffer dough and cut or pinch off pieces and make biscuits, or you can add less flour and make hotcakes.

"Take the remaining sour dough and mix it with a little more flour, sugar, salt and water and set aside for the next baking."

Mrs. Joe Hoffman
Atchison, Kan.

Egg Butter or Backwoods Preserves

When fruits for preserves were scarce in early days, many a

family enjoyed Egg Butter. My grandma lived to be one hundred years old. This is her recipe.

"Bring to boil one pint of sorghum. Have ready two well-beaten eggs. Add a bit of nutmeg or cinnamon to the sorghum and stir in the eggs, stirring constantly until thickened, which won't be long. As it boils up, add a small pinch of soda as it will cut the too strong taste of the sorghum."

Mrs. Ed Jarvis
1030 Lawrence St.
Topeka, Kan.

Lemon Crackers

How well I remember my mother making Lemon Crackers with the ammonia melting in a tin cup on the back of the stove! Mother's recipe was lost, but the memory of those delicious crackers did not fade. I was unsuccessful in trying to find a recipe until a relative in my native West Virginia sent one. It was not very specific and, among other things, called for 2 1/2¢ worth of ammonia. How much was that three generations ago?

I asked the Home Economics Department of the State University to modernize the recipe for me, but these experts could give me no help. Years later, in a recipe contest, one of the entries was for Lemon Crackers! Here's the way it goes:

"Mix together one cup lard, two cups sugar, two eggs, two cups sweet milk, one teaspoon oil of lemon, one-half ounce carbonate ammonia, pulverized and dissolved in a little warm water. Add enough flour to make a stiff dough. Knead well, roll very thin, cut into squares, prick with a fork and bake quickly in a hot oven. Let cool before serving. This makes about five pounds of crackers."

Salt was not mentioned in either recipe. Guess they just took that for granted.

Mrs. Paul Murdoch
4211 Holdrege St.
Lincoln 3, Neb.

Baking Powder (Reprinted from Capper's Weekly)

This recipe for baking powder or "Fast Yeast Powder" was taken from a cook book revised in 1870.

"Sift together two cups cream of tartar, one cup of baking soda, one cup flour and one cup sugar. Store in airtight containers and keep in a dry place. Use one teaspoon for every two cups of flour."

Lucille Kelly
R. 2
Ewing, Mo.

The 'Ley' Barrel (Reprinted from Capper's Weekly)

Grandmother not only had to make her soap, first she and Grandpa had to make the lye. In this old recipe for making it, it was spelled "ley."

"Bore a hole in the bottom of a barrel; place an earthen vessel under the hole to catch the ley; place straw in the bottom of the barrel using ten bushels of ashes and one bushel of unslacked lime. (Ashes should be kept dry and they should be strong ashes—hickory ashes are best.)

"Pour into the barrel one layer of ashes and on top of that sprinkle lime and add water (rain water is best). Continue alternating the ashes and lime and adding water until all the ashes and lime are used up.

"The ley is ready as soon as it is strong enough to bear up an egg with only the top visible above the water. The top area should be the size of a dime above the ley. If the ley is not strong enough, keep pouring the ley back through the ashes and adding more water as necessary."

Lucille Kelly
Ewing, Mo.

To Remove Wrinkles (Reprinted from Capper's Weekly)

(If my mother-in-law were living, she would be 92. This recipe was among her things.)

"Melt together one ounce white wax, two ounces strained honey and two ounces of the juice of lily bulbs. The foregoing melted and stirred together will remove wrinkles."

Mrs. Albert W. Snyder
Box 701
Maxwell, Neb.

Starch

"Wash a peck of good wheat and pick it very clean. Put it in a tub and cover it with water. It must be kept in the sun and the water changed every day or it will smell very offensively. When the wheat becomes quite soft, it must be well rubbed in the hands and the husks thrown in another tub. Let this white substance settle and then pour off the water. Put on fresh and stir it up well and let it subside. Do this every day till the water comes off clear, then pour it off. Collect the starch in a bag, tie it up tight and set it in the sun a few days. Then open it and dry the starch on dishes."

Mrs. C. C. Dosien
Valley Center, Kan.

Clover Blossom Vinegar (Reprinted from Capper's Weekly)

Clover blossoms sometimes were used with hop yeast to make vinegar. This is from my collection of old recipes.

"Pick the clover blossoms without stems or leaves attached. Place a peck of the blossoms in a stone jar, add two quarts of brown sugar, one quart of molasses and four and a half gallons of boiling water. When this cools, add one and one-half pints hop yeast. Mix the ingredients well. Place a white cloth over the jar and let stand in a cool place for 14 days. Strain the vinegar off and seal it in jars."

Mrs. Lorin Best
R. 1, Box 128A
Mission, Texas

Face Powder (Reprinted from Capper's Weekly)

"Take one-fourth pound of wheat starch pounded fine. Sift it thru a fine sieve or a piece of lace. Add to it eight drops of oil of rose, oil of lemon 30 drops, oil of bergamot 15 drops. Rub together thoroly."

Mrs. L. A. Houston
Rockville, Neb.

Perfumery (Reprinted from Capper's Weekly)

(This recipe was taken from an old scrapbook handed down in my husband's family and containing clippings dated between 1816 and 1878.)

"Spread fresh, unsalted butter on two plates of the same size. Then fill one plate with roses, jasmine, violets or any flowers you wish. Turn the other plate over it and let stand for 24 hours. Then scrape off the butter from the plate and put in some alcohol. Cork tightly."

Mrs. Dean Herlacker
Farlington, Kan.

How To Grow Thin (Reprinted from Capper's Weekly)

(I thought dieting was a strictly modern concern until I found this hint in digging thru old cook books.)

"Drink as little as you can get along with comfortably, no hot drinks, no soup, no beer and only milk enough to color the luke warm tea or coffee you drink. Eat chiefly stale bread, lean meat with such vegetables as peas, beans, lettuce, in moderation. Avoid watery vegetables such as cabbage, potatoes, turnips, etc. No pastry whatever. Limit yourself to seven hours of sleep out of the 24, and take plenty of exercise in the open air."

Mrs. Roy Dickey
6943 Walrond
Kansas City 30, Mo.

For Spring Fever (Reprinted from Capper's Weelkly)

"For Curing Spring Laggins" is the interesting title for this recipe I found in my mother's "receipt" book, as it was called in those days.

"Take sarsaparilla, yellow dock, black alder bark, burdock root, sassafras bark, wintergreen. Of each one ounce mixed with four pints of sirup. A wine glass of this three times a day."

Mrs. E. A. Stowell
Boscobel, Wis.

Salve (Reprinted from Capper's Weekly)

This recipe for salve has been in our family for at least 85 years.

"Melt together one-half pound lard, one-fourth pound rosin, one-fourth pound beeswax and one ounce gum of camphor. After it is taken from the stove, add a little turpentine when almost cold."

Mrs. F. W. Parrish
1706 Arizona Ave.
Santa Monica, Calif.

Grandma's Eyewash (Reprinted from Capper's Weekly)

This is Grandmother's recipe for eyewash that she claimed to be "the best eye water ever made for man or beast."

"Take three fresh eggs and break them into one quart of clear, cold rain water; stir until thoroughly mixed; bring to a boil on a slow fire, stirring often. Then add half an ounce of sulphate of zinc (white vitriol); continue boiling for two minutes, then set it off the fire. Take the curd that settles at the bottom of this and apply to the eye at night with a bandage. It will speedily draw out all fever and soreness. Strain the liquid through a cloth and use for bathing the eyes occasionally."

Mrs. Ambrose Fischer
Strawberry Point, Iowa

Cocklebur Cough Sirup

"Gather cockleburs as clean as possible. They can be rinsed off by pouring hot water thru them. Put burrs in a heavy cloth sack and boil long enough to get the strength out of them. Strain juice several times to be sure there are no stickers in it. Add sugar to make pretty good and sweet and boil down into a sirup. Use for colds and coughs—even whooping cough. It may make you feel a bit feverish at first, but that wears off."

Mrs. Wendell Taylor
Luray, Kan.

For Whooping Cough

In the days when whooping cough was common and children whooped for weeks or even months afterward, this recipe was considered highly valuable—even in our family of teetotalers.

"Take equal parts of strained honey, olive oil and whiskey and mix well. Give a teaspoonful three times a day."

Louise Fowler Roote
1025 Taylor
Topeka, Kan.

English Plum Pudding

My pioneer mother's cook book still opens automatically to her recipe for Soft Gingerbread, a childhood favorite of ours. But since the gingerbread is not very different from the way we make it today, I'm going to give the English Plum Pudding recipe from the same cook book.

"Take six ounces of suet, mind you skin it and cut it up fine. Just you use the same quantity of raisins, taking out the stones, and the same of currants. Always wash your currants and dry them in a cloth. Have a stale loaf of bread and crumble, say three ounces of it. You will want about the same of sifted flour. Break three eggs, yolks and all, but don't beat them much. Have a teaspoon of ground cinnamon and grate half a nutmeg. Don't forget a teaspoon of salt. You will require with all this a half pint of milk and four ounces of white sugar. In the old days angelica root candied was used; it's gone out of fashion now. Put that in—if you have it—not a big piece and slice it thin. You can't do well without a half ounce of candied citron. Now mix all this up together, adding the milk last in which you put half a glass of brandy. Take a piece of linen, big enough to double over, put it in boiling water, squeeze out all the water and flour it; turn out your mixture in that cloth, and tie up tight; good cooks sew up their pudding bags. It can't be squeezed too much, for a loosely tied pudding is a soggy thing, because it won't cook dry. Put in five quarts of boiling water, and let it boil six hours steady, covering it up. watch it and if the water gives out, add more boiling water. This is a real English plum pudding with no nonsense about it."

Louise Fowler Roote
1025 Taylor
Topeka, Kan.

Pea Hull Soup

I am 83 years old and this recipe for soup made from pea hulls belonged to my mother. Pioneers really knew what "Waste not, want not" meant.

"After hulling the peas from the pods, tie the pods in a bag made of some coarse cotton or linen cloth; place the bag in cold water over a moderate fire; boil until the sweetness is extracted from the pods; remove the bag; then squeeze it so that all the juice will be left in the water; after that is done, season as any other soup, and depend on it you will have as fine and highly flavored a plate of soup as ever graced a table or tempted an appetite. Don't throw away the pods of the peas when a soup can be had at so small a cost."

Mrs. Isabelle Quinn
411 Euclid Ave.
Cherokee, Iowa

CHAPTER 16: The Gold Rush

Pancake Pilferer

My grandfather had crossed the plains to California with the 49ers, and was a great teller of stories of the Gold Rush days. After supper, we children used to sit on Grandpa's old buffalo hide laprobe while he spun amazing stories of the perilous trek across the plains. One of the stories he told was of the mystery of the disappearing pancakes.

Every day the wagon train was beset by Indians. Not the fighting, scalping, murdering kind—but more the "trick or treat" kind. As soon as camp was set up, these visitors would be around—poking into things, prying, begging and stealing.

One morning the cook, a big fellow, had prepared a roaring fire and was baking pancakes for the whole lot of travelers, turning out a dozen at a time. As usual, an Indian stood close to the fire. Apparently he was too proud to watch proceedings, for he kept his back turned and his blanket drawn closely about his burly frame.

Soon the puzzled cook began to miss pancakes. As he would turn to bake a batch, the ones on the platter disappeared as tho into thin air. In exasperation, he stopped baking and scratched his head. He had it!

He placed the fresh-baked cakes on the platter and turned to the fire, but instead of attending to the baking, he cautiously turned back, took a firm grip on the handle of the red-hot skillet with one hand and quickly lifted the big buck's blanket from behind, making connection with bare hide.

There was a whoop that split the heavens! The Indian's arm shot up and outward, the blanket sailed thru the air, and pancakes rained down in all directions on the startled 49ers.

The Indian took off across the prairie. No doubt he wore that pancake brand for the rest of his life.

Mrs. C. O. Barnes
15 University Place
Redlands, Calif.

A Gold Seeker's 13 Letters

Miss Anna Ryan Greenwell of Lakenan, Mo., has a packet of 13 letters over a century old written by her grandfather, Alden Rice Grout, as he made his way across the plains to the gold diggings in California and labored there hoping to make a fortune. The letters were written on thin blue and white paper. When the writer had covered the pages, he turned to the margins and filled them with last minute thoughts. One space in the center was left for the address, and the page was sealed with a dab of sealing wax. Here are excerpts from the letters:

"Richmond, Mo., April 22, 1849—If practicable, I wish to go on to California as I have been foolish enough to start, and try to be paid for some of the trouble and anxiety I have already felt by leaving a pleasant home for the miserable life I am now enduring."

There was a disagreement among the leaders of the 32-wagon train. At Ft. Laramie, Wyo., Mr. Grout and seven other men broke away to form their own company. He wrote: "The captain, as good and clever a man as ever lived, wished to please all; the consequence was nobody was pleased."

Estimating that they had passed 1,000 dead oxen in 150 miles of alkali plains along with deserted wagons, tools and clothing, he says: "I can think of nothing I ever read to compare with it except Bonaparte's excursion to Russia."

The company arrived at the diggings near Sacramento on

Sept. 16, 1849. Mr. Grout's letters soon were filled with admonitions to his friends in Missouri to remain there. There were days when he dug $200 worth of gold from the hills, but these days were preceded by weeks of labor that netted nothing. To a friend asking his advice about making the trip, he wrote: "I doubt not but you almost weekly hear of this one and that of having taken out his pounds and pounds of gold. But do you hear of the one thousand and one that have died, or are lying sick, unable to labor for their bread and not any means to buy it with?"

The fantastic prices took their share of his hard-earned gold. At one time he wrote of flour being $2 a pound; onions, $1.50; potatoes, $1. Hay was $10 for 100 pounds, and cornmeal $25 a barrel.

In all his letters, Mr. Grout begged for word from home. His wife wrote, but it was May 29, 1850, more than a year after he left home, before he received his first letter. It had been sent east to New York and by steamer around South America.

The last letter in the packet informs his wife he will leave for home in December, 1850. He had found gold, but not enough to keep him in the treacherous golden land. There was much, he discovered, that gold could not buy—the love of wife and children, peace of mind, a little comfort, such a small thing as a table on which to write.

"I have been reading this letter over," he laments, "and confess it goes a good deal like riding in a lumber wagon over frozen roads."

Mrs. Robert G. Lanham
Monroe City, Mo.

'Commencement' Party

Not all the story of the Gold Rush was one of grief and disappointment. There was romance, too!

Cora Reed's husband died two weeks before the wagons started west. She didn't know what to do. If she stayed she would be all alone because her daughter and her family were part of the party going in search of treasure. Finally, they persuaded her to go with them.

She had a good team and wagon and insisted that she would drive it herself because she was used to animals. By the time they had reached Leavenworth, Job Witherspoon, a widower, had appointed himself as Cora's guardian and he helped her with her team in all the hard places. Long before they had reached the gold fields, they had decided to marry.

"But, Ma, you don't know much about him," objected Cora's daughter.

"I know that any man as nice to young-uns as that man is will be good to a wife," Cora replied.

"Pa, how in the world will you find work enough to take care of a woman?" Job's daughter asked.

"I've noticed," said Job, "that if a body isn't a'feered of work, they generally can find something to do!"

When they realized how determined Job and Cora were, all the members of the wagon train decided to give them a "commencement party." That's not what is sounds like. It was the name for a pioneer wedding shower!

Each family checked and rechecked possessions to see what they could find to give the bridal couple. There was much speculation as to whether or not there would be a preacher to do the marrying at the mining camp. Some trappers passing on the trail stopped to visit and assured them that the camp had a preacher, one who had been quoted as saying, "There's enough sin in this camp to keep one preacher busy the rest of his life."

Cora Reed was a very hospitable woman and it disturbed her that she couldn't have plenty of refreshments for the friends who were giving the commencement party for her. It was near the end of the journey and supplies were low. She knew she didn't have materials to bake cakes.

All of the women got together and pooled their supplies. The dried apples they had brought from home came in handy. They chopped and cooked them and used them in a sort of spiced cookie. By scraping together all the supplies they could possibly spare, they figured they would have enough to make two cookies for each person at the commencement party.

That evening there was a good bit of "joshing" and then they sang and danced. Each person had concocted a gift of some kind. One woman made an apron from some material she had brought with her. Cora had started out with her best dishes, but her wagon had toppled over on a rough stretch of trail and every dish had been broken or cracked. Other women gave her some of their dishes for wedding presents.

The men got together and divided their tools with Job because he was handy with them and doing carpenter work might help him support a wife when they got to their new home. It was amazing to see the gifts that came from the various wagons and what a good showing they made. The commencement party gave Job and Cora a rosy start into their new adventure.

Leona Haskell McDaniel
Topeka, Kan.

A 49er at Sixteen

When my grandfather was 16 years old, in 1849, he traveled with friends to search for gold on the West Coast. Their 15-wagon caravan was on the road six months from Illinois to Sacramento, California. The women, children and provisions used up most of the space in the wagons, so the men and boys walked many miles of the trip.

A young man among the group walking was a "smartie." He noticed a young squaw fishing on the bank of a stream. He was carrying a rifle, and on impulse he shot the squaw. Near sunset, the Indian warriors surrounded the caravan and demanded the killer. Unless he was surrendered, all whites would die. The young man was scalped before the very eyes of his friends and left to die on the trail.

Sacramento was just a tent city when they arrived, with not a woman in the town. Thirty dollars worth of gold was considered a good day's diggings. But prices were high. A pair of chickens sold for $12; oysters were $5 a pint. My grandfather lived in California 16 years, and during that time went to Cuba for a short time. When he returned to Illinois, he had a total saving of $800.

Mrs. Fred Scar
Earlham, Iowa

Murdered at a Wagon Stop

When word of the great gold bonanza in California reached the South, my great-grandparents were living in Mississippi. It was decided in a family conference that the father would go West in search of a fortune and a new home and the mother and children would live with her parents until he could come for them.

He started out in a brand-new wagon loaded with supplies, including picks and shovels, and pulled by a fine team of matched black horses. He sent home letters as he pushed westward, but one letter from a settlement in Wyoming proved to be the last word ever heard from him.

After several months, his father and brother went over the same trail trying to find him. At a wagon stop and inn in Wyoming a terrible scandal had been uncovered. Countless travelers had disappeared there, and bodies had been found in shallow graves near the inn. A mob had been robbing and murdering "guests" from wagon trains and several of the scoundrels had been hanged for the murders. His relatives lost hope when great-grandfather's team of matched blacks was found at the inn.

Mrs. Tom Tyre
R. 2
Miami, Mo.

Food Was Expensive

During the gay 90s, we lived 20 miles west of Denver. I have seen as many as a dozen covered wagons going thru our farm.

Some of the people were gold hunters who settled in the mining camps of Idaho Springs, Leadville and others. Food was high in the mining camps—potatoes were 24 cents a pound; butter was $1.25 a pound; eggs were $1.25 a dozen; flour was $25 a hundred; and hay was $250 a ton.

Emil Rudin
Stuart Draft, Va.

Gold Dust for Flour

My aunt and her twin brother were great-grandniece and nephew of President Andrew Jackson. My aunt lived to the age of 104. They were born in Ohio in 1842 and moved to Illinois when they were quite young. There my uncle learned to set type when he was so young he had to stand on a box to reach the type case.

During the mad Pikes Peak gold rush in 1863, when everything in Colorado was paid for in gold dust, my aunt, a small woman who weighed scarcely 100 pounds, drove an ox cart across the plains from Illinois to Denver carrying a sick husband and a baby. She walked a good share of the way, leading the ox team.

This ox cart was a part of the Barber Train. Tho they never were attacked by Indians, they came across two still-burning ruins of trains just ahead of them and saw the dead bodies of fellow pioneers.

Denver, at that time, consisted of only a few scattered buildings and tents. My aunt started a tavern and one of her boarders was the Ute Chief Colorow. Her husband freighted across the plains from the Missouri River until the railroad took over in 1871.

My uncle bought a load of flour at St. Louis and hauled it with a team of oxen. When he arrived at Denver, the town was out of flour and there was a mad rush to buy it. Some was paid for in gold dust that had to be weighed. When he quit freighting he helped print the first edition of the Rocky Mountain News.

These pioneer twins were small people and they worked hard,

but they outlived many people who were bigger and probably had a much easier life.

Mrs. Amy C. Shaw
Mound City, Kan.

Matilda's Three Marriages

"Tilda, I wish you would quit egging your pa to go to the Colorado gold fields. You know he has an itchin' foot and is only too anxious to have an excuse to move on again. When his folks died and left him the home place in Ohio all stocked, I thought maybe he would stay put for a spell. But it wasn't long until he said things were gettin' too civilized and we moved to Indiany. We just got settled and he wanted to push on farther and we came to Wisconsin. The worst of it is—every move we make we get poorer."

"Ma, I feel bad about it, but John is just bound that we are going. I don't feel a rough mining camp is any place for two little boys and I know the baby is too young to take. He says since I'm nursing her she'll be all right. I've tried telling him my milk will go back on me, but he thinks I'm just making it up. Did you ever see bluer eyes than this girl has, Ma?"

Susan looked at her eldest daughter, a beautiful young woman with a nice figure, quantities of red-bronze hair, dark blue eyes and a peach-blow complexion. She had been a merry, light-hearted girl until she had married John.

"I don't know, Tilda, I'll think and pray about going. I do wish John would quit his gambling and drinking."

Matilda had gone for a visit with her Aunt Rach, who was very romantic and had persuaded her to marry John. It had been a sorry marriage because John was cruel when he drank.

The next morning John came in and said to Tilda, "Well, old woman, get ready! I just bought a yoke of oxen for $100. We'll go next week so you'd better get things ready. Don't take anything you can get along without—some of these days you'll be wearing silks and satins!"

Matilda had a feeling of premonition of some impending disaster that hung over her like a dark curtain, but she was relieved to learn that her parents had made up their minds to go on the gold hunt, too. The little boys were all excitement and would lean over the crib and tell their baby sister all the wonderful things that would happen.

It was a beautiful May morning and everything looked green and fresh. The birds were singing and wild flowers were beginning to appear. Matilda looked over her little house and at her belongings and then would go out and measure the wagon again and wonder how she could ever get in all the things she needed. No matter what she took she'd always remember something they had left and needed. Among the things she took were two feather beds piled on top of each other and a leather-covered trunk with clothes and a few choice possessions. I think all the women slipped in a few things that didn't come to light until they arrived in Colorado. Susan had a quantity of dried apples left over and she took them and they were all glad she did.

The trip to Kansas was not so bad, and the travelers all felt encouraged because they had learned a good bit that surely would help them on the wilder part of the journey. They arrived at Leavenworth late in the evening and were amazed at the wagons. They hadn't dreamed so many people were going to Colorado and Oregon. They decided to take the Oregon Trail and cut down from Ogallaha to Denver.

In the camp that night they heard something that made their blood run cold. It seemed that on the last trip out a man had said he would shoot the first Indian he saw. No one took him seriously, but in about the wildest part of the trip they saw an Indian girl sitting on a log watching a train go by. They heard a shot, and to their horror they saw the girl topple over. The men were aghast because they knew there would be reprisals.

At daybreak the camp was surrounded with Indians who demanded to know which man had shot the chief's daughter. The leader of the train knew he had to give him up or everyone would be killed. The Indians lined the whites up and everyone had to stand and watch while the guilty white man was skinned alive! Children screamed and women fainted. The suffering man crawled down the bank for water. He lived three hours. It wasn't a tale to comfort nervous women, and some just refused to go on.

Matilda was more concerned about her baby than about Indians, tho, as they started out across Kansas. The infant girl had summer complaint, and the medicine they had bought at the post didn't help much. The third day out the baby was much worse, and Matilda was beside herself. As they made camp that night, the child died. There was no material with which to make a coffin, so Matilda took her new Paisley shawl out of the trunk and wrapped the little body in it. They read a psalm, had a prayer and sang a hymn. The next morning the wagon train was driven over the grave, so the coyotes wouldn't find it.

One of the travelers was Old Jeb, a man said to be so lazy he had callouses on his hunkers. But before the trip was finished the leader said he was one of the most valuable men they had. It was getting hotter and drier all the time and the winds were scorching. The children were tired and cross. The mothers were nervous and fidgety and their eyes burned and their lips were cracked and parched with alkali dust. It was Old Jeb who walked among the wagons and told funny stories and cheered up the dispirited. He took a pan of water and a rag and went down the wagon train and washed the nostrils of the oxen. The leader said of him, "Old Jeb is good for man and beast."

Blood-curdling yells awoke the travelers one night. Matilda put both boys under the two featherbeds and told them if they so much as peeked out she would skin them alive. Just then there was a zing—and an arrow was embedded in the trunk. It was plain that the Indians were trying to stampede the horses, and John brought his beautiful black horse to Matilda and told her to hold and quiet him. Not 20 feet from her wagon, Matilda saw an Indian shot from his horse. At daybreak the Indians left as suddenly as they had come. Several whites had been injured, but none fatally. When the men examined the body of the "dead Indian" they discovered he was a white man wearing a war bonnet. They knew then that they had dealt with brigands after horses.

The wagon train made it thru to Colorado, and John threw up a cabin near a Boulder gold mine. The miners were a wild bunch and Matilda said she was more afraid of them than of the Indians.

It was a discouraging year. John drank and gambled most of the time, and Matilda's father was not adapted to work in a mining camp, so he did carpenter work for others. At the end of the year, Susan and Sam went back to Wisconsin. Matilda felt desolate at seeing them go, but she was glad they still had money left to get back.

John's drinking and gambling got worse and worse and he was cruel to Matilda and the children. Finally, she could stand it no longer and separated from him. It was very hard for a woman to make a living in a mining camp, and Matilda did menial work to make a little money. She washed and cooked for the miners and even panned a little gold. John stayed around camp just to make life miserable for Matilda, it seemed.

The next year the still-young Matilda was courted by a man of some means whose wife had died the year before. He was kind and thoughtful, a very different type than John. Hannibal was fond of the little boys and they adored him. Matilda married Hannibal and began to be happy again.

Hannibal thought it would be fine if the boys had a pair of cats for pets. He ordered a pair sent by the next train, but they were so long coming that when they arrived they had five kittens! That night three of the toughest men in camp, Blackie, Swede and Charley, knocked on the door and said they wanted to see "them cats." The boys were excited and anxious to show their pets. One kitten had a bad eye and was dubbed One-Eyed Pete for a man in camp. Another kitten was named Liz for a woman camp follower who could shoot, swear and drink with the toughest. The rough miners were entranced with the kittens and with little Henry and Johnny.

Matilda and Hannibal decided to give the children their first real Christmas that year. They invited the boys' three tough miner friends. For the occasion, Blackie shot a nice fat deer and Matilda made venison mincemeat with some of the dried apples they had brought from Wisconsin. Before Christmas day, the men made a trip to Denver and bought gifts for the boys and a new Paisley shawl for Matilda. They had heard how she had buried her baby in her shawl. The three men had a wonderful time shopping and spent all their gold dust so they didn't have any left for a "bender." They had a wonderful Christmas day and one of the men said to Matilda, "I declare, Missus, you're gettin' prettier every day!" And she was; she didn't have the worry she had had before.

The first warm day of spring Matilda went to see a sick woman and left the little boys in the yard to play. When she came out, the boys were gone. Matilda was frantic and began to search all over town for them. Then a miner told her he had seen John with the boys and the father had boasted he was stealing the boys and taking them to California where they would be bound out. As long as Matilda had a hard time taking care of the children it was all right with John. When things became easier for her, his jealous nature couldn't take it. He didn't really want the children; he just wanted to ruin Matilda's new-found happiness.

Matilda had left Wisconsin with three children. Now her arms were empty. She was heartbroken. The three tough miners to whom the little boys had endeared themselves wanted to do something. But what could they do? John was the children's father. It's a good thing, tho, that the men couldn't get their hands on him!

Black measles hit the camp in its most virulent form. Liz, the camp follower, was delirious and no one would go near her. Matilda told Hannibal she was going, and he went with her. They cared for Liz the best they could, but she died the next day. They asked the deacon to say a few words at her funeral, but he said she was damned to hellfire and he wouldn't go near. When the little group was gathered at the grave, it was Matilda who said, "Well, folks, you know—or ought to know—the Lord's Prayer. Now we're going to say it!" A few days later Matilda came down with the disease and nearly died.

Hannibal at last achieved the thing he had been working toward—he had raised enough gold to buy Henry and Johnny out of bondage! He arranged that the boys should be delivered to Matilda's old home in Wisconsin and that Matilda should go back to visit her relatives and meet the boys.

In the meantime, Hannibal planned to go on a spring prospecting trip with a party going far back into the mountains. Two days before Matilda was to start to Wisconsin, John returned from California and said that he, too, was going on the prospecting trip. Matilda begged Hannibal to change his plans and not go. "The worst Indian is white beside John, and I'm afraid for you, Hannibal," she said. But her husband said he'd always taken care of himself and refused to change his plans.

Long before Matilda reached Wisconsin she knew she was to become a mother again. She arrived at her old home, but months and months went by and there was no word from her husband, nor did the boys arrive. One week before their blue-eyed sister was born, Henry and Johnny arrived safely. But the joy of their homecoming was dimmed by the news brought by the man who had delivered the boys. He said that two stories about ill fate to the prospectors were going the rounds. One was that Ute Indians had killed them and the other was that a band of robbers had attacked them for their gold.

Again Matilda had to provide a living for her family. As soon as she could wean the baby, she walked the 16 miles to a town where she could get work in a hotel, doing as much work in a day as two women should have done.

She hoped against hope that Hannibal would walk in some day. But it was John who walked in! Then one night at a party, John told Matilda's sister that Matilda never would see Hannibal again because his body was at the bottom of a pit—and he showed her Hannibal's watch to prove it!

Because Matilda would have nothing to do with John, he was furious. At a neighborhood dance, he brought in a sack of candy and offered a piece to Matilda. She said she didn't want any, but when it looked as if he might make a scene, she took a piece. The family dog strolled in and she gave the candy to the dog. In a few minutes the dog died of convulsions!

Matilda worked at the hotel for three years and scarcely saw her children, who stayed with their grandmother. Finally, she married a man named Jim in the hope of having a home for her boys and little girl. She was scarcely established when John stole the boys again and took them back to California. She never saw little Johnny again, but Henry came back when he was 17 to see his mother and made many trips back as long as she lived.

Matilda bore nine children during the years of her third marriage, but the marriage was unhappy and she became embittered as she grew older. In her early 50s she had pneumonia from which she did not recover. Not many of the 13 children she had borne were with her at the last.

My mother was the little girl born to Matilda and Hannibal after Matilda returned to Wisconsin. I have the brooch that was Hannibal's wedding gift to Matilda. When I was a child I saw the little leather-bound trunk that Grandmother Matilda took with her on her wagon trip west. And the Indian arrow was still in it!

Leona Haskell McDaniel
433 Lincoln
Topeka, Kan.

CHAPTER 17: Land Races Into Oklahoma

Eva Made the Great Run

Her dimming eyes sparkle with excitement when Eva Wintermute talks about the Cherokee Strip Run in 1893. With a keen memory for detail, she adds new zest to the story when she begins:

"All we heard during a long, hot, miserable summer was talk of the Big Run. The Cherokee Outlet was the last frontier, a land of lush grass and fresh streams–five million acres of 'Promised Land.'

"I was only 17, too young to get a claim, but I had promised to marry Walter the next spring, so we looked forward to getting a claim of our own. Since Father's death I had helped Mother keep the farm going and could handle a team like a man, so I agreed to drive the supply wagon for the Wintermute brothers, who would make the Run on horseback.

"Mother and I set out in a covered wagon from our home near Chautauqua Springs about a week before the Run and made the hot, wearisome trip to Arkansas City, where the men planned to register.

"We camped in the yard of a friend's home and cooked meals to take to Walter and his brothers, while they stood in line three days to register. Between meals we fanned and shooed flies and visited with camp neighbors. Thirty thousand people were crowded into that town of 4,000.

"All kinds were there, the peddlers, the fakers, the blind and the greedy. Plain folks, whimpering children and yelping dogs, horse traders with business, young men selling water at ten cents a cup. The heat was fierce and a thick dust was on everything, swirling in the streets, clinging to our sweaty faces. Nothing but Strip talk was heard, but the most urgent need was for a cup of fresh water.

"On the day of the Run, Mother drove our wagon and I took the supply wagon, accompanied by Art Skaggs, a 12-year-old boy hired by the Wintermutes. We wedged into the line 12 miles west of Arkansas City. As far as the eye could see were covered wagons, buggies, buckboards, carts, bicycles, even a few surries. Thousands of men on horseback, riding Texas cow ponies, common horses, mules or racing thorobreds, shipped in from the East.

"The suspense was electric when the starting signal was heard at noon. The horsemen vanished in clouds of dust. I was doing fine until we met a prairie fire and the team bolted. Racing across gullies and over the hills, the wagon swayed so I thought we would turn over any minute. It was everybody for himself, and no racing horseman would stop to offer help. Art and I used all our strength pushing the wheel brakes and finally dragged the wheels enough to slow the team.

"We camped that night on the Chikaskia River near Blackwell, and soon after dark a wagon drove into camp and Mother stepped down.

" 'I got a good claim, but just couldn't stay there after dark by myself,' she said. The next day we located the Wintermutes on the Salt Fork River near Tonkawa. Walter's claim turned out to be on school land, which had to be bought!

"So we started the trek back to Kansas, forgetting our dreams of starting a new home on free land."

Jessy Mae Coker
Sedan, Kan.

Traded Team for a Claim

My father went alone to make the Run into Oklahoma, but because of high water he was one day late. It was a bitter disappointment, but he was not a man to give up easily. He found a man who had made the Run, but who already was sick of his bargain and willing to trade his claim for Father's team of horses. The trade was made and Father soon was on his way back to Nebraska for our family.

We covered about 25 miles a day with our three wagons, five cows, nine horses and six children. I was nine years old and rode my pony all the way as it was my job to drive the cattle.

When we arrived at the Cimarron River it was running full and, of course, there were no bridges. Father rode my pony across to see how deep the water was before we started with the wagons. We were afraid, but we fastened the wagon beds down so they wouldn't float away and the crossing was made successfully.

When we came to Dover, the officers demanded our horses for the use of a posse formed to catch the bandits who had just robbed the train at Dover. It was three hours before our horses were returned to us. The bandits were not caught.

We spent the night in Kingfisher and got supplies and lumber for our new home. The next morning we started out across the prairie. We had only Indian trails to follow the 18 miles to our homestead. It was desolate country as we pulled onto our claim.

Before morning Mother was frantic because we could hear the tom-toms from an Indian camp in the distance and nearby we could hear the howl of a coyote. Mother begged Father to turn back. It was no place to raise a family, she said. But we stayed! It was hard those first years, but eventually times improved and our homestead became a true home. Young folks of today will never have experiences like these to remember.

Roy Shaffer
Watonga, Okla.

Raced on Horseback

When I was about nine years old, the Cherokee Strip in Oklahoma was being opened for settlement. Most old folks will remember that great race for land. Days and weeks before the date, people on horseback, in buggies, covered wagons and vehicles of every sort were camped on the Kansas border to make ready for the race.

My father and my oldest sister's husband decided to leave our homestead in northern Nebraska and make the race. They took a horse each and started for the Kansas border.

They left my mother, my oldest sister, a grown brother and six younger children to pack and bring the belongings if they won some land. A few days after the race, we received word that Father had won a tract of 160 acres.

There was a lot of scrambling around. We had raised our own wheat and had flour ground at the mill. One wagon was loaded mostly with 100-pound sacks of flour. My sister, about 20, drove one wagon. It was slow traveling and we averaged about 20 miles a day.

We'd never seen apples growing on trees and when we passed orchard country we thought we were in the land of promise. I remember how we spread a cloth on the ground to eat, and we kids sat on the wagon tongue with our food.

After a long, hard trip we landed in Blackwell, Okla., where my father had bought some town lots and built a small house while we were on the road. We lived there until he had time to make a dugout on the land he had won in the race.

Mrs. J. L. Veatch
1602 Trinidad
Dalhart, Texas

Awakened by Gunshot

When my father decided to pioneer in the Oklahoma Territory, he was a farmer in Nebraska. He sold his farm and started on the long trek to the new land with his family of nine children.

When we arrived at Caldwell, Kan., we heard that the government had ordered all cattle taken out of the Cherokee Strip and the flies—deprived of the cattle to feed on—were simply eating the horses alive. Mother took her bed sheets and made covers for the horses, and then Father drove them right into the swarms of starving flies.

The horses couldn't be held back. They went at a fast trot and couldn't stand still long enough to drink water. Father kept going until dark stopped the flies. We made camp, but not for long.

We were awakened by many gunshots. Father supposed it was horse thieves, so we loaded up and started on. When we arrived at Kingfisher we learned that the shots were those of train robbers. That was the night the Dalton boys were trying to rob the Rock Island train near where Enid is now. That was my first covered wagon experience. The second was more thrilling.

I married a young man who was pioneering farther west. We loaded our belongings in a covered wagon and started to Washita county. On the way the wagon tongue broke and there was no way to get it repaired. We were crossing an unsettled part of the Oklahoma Territory.

The only way we could keep going was for me to keep a whip in my hand as my husband drove. Whenever one horse got a little behind, I had to tap him on the rump to make him step up. We got nearly to Arapaho by night, but our troubles were not over. We had to ford the Washita River. The horses were so tired and the river bank so slick that they couldn't pull the wagon out of the river. We had to go and get a farmer to help. When we were only half a mile from our new place, the horses failed again. We just unloaded our wagon and slept by the creek until morning. Then my husband went to get a neighbor to help again. But all's well that ends well, and this place has been my home for 56 years.

Mrs. Clara Grubb
420 N. Market
Cordell, Okla.

Trip in A Surrey

We started from Missouri for Kiowa, Kan., to be ready to make the race to take up land in the Strip.

Someone had given me a canary and the cage was hanging in

the surrey with the fringe on top in which Mother, Grandmother and I were traveling. We got behind and our mule was tough-mouthed and in a hurry to catch up with the crowd. The bottom of the cage fell out and the bird flew away into the bushes. I couldn't catch it and we had to go on.

We were two or three weeks on the road, and I saw Indians for the first time. Some of the men and boys wore their hair in long braids. We rented a house in Kiowa and waited for the opening. I remember many trips we took looking over the Strip. There were pretty wild flowers and I picked up buffalo horns until I had almost a wagon bed full.

I remember a celebration—Fourth of July, I guess when they chased greased pigs. The one thing that stands out in my memory was a race of cowboys down the main street. Several geese were hung on a wire stretched across the street just in reach of a man on horseback. The trick was to pull off a goose head as the horses ran under the geese. One of the fellows jumped from his horse and gathered a handful of sand and got a goose head on his next try. I remember visiting a big ranch where the family had a nice chair made of buffalo horns.

Ethel McCurdy
R. 5, Box 490
Springfield, Mo.

Widow Saved Her Claim

I am 88 years old, and my husband made the run into Oklahoma for a farm.

We moved to the place in October with a five-week-old baby and two other children. My husband drove the covered wagon and I drove the covered spring wagon. Before we started I baked up a lot of bread and some cakes and packed them in a wash boiler. One evening one of the horses got loose and ate nearly all of our bread and cakes!

Our home was to be about one and a half miles across the Cimarron. It didn't look like much. Our first home was a dugout, then a soddy, then a log house. Finally, we built a nice home.

In 1894 a little girl was born, making four children. In July that same year my husband died of cancer. Things sure looked gloomy, but the children and I stayed on. I had to prove up on the farm the next year to keep from being contested. I have a patent signed by Teddy Roosevelt.

When my husband died the oldest child was ten and the youngest three months. I plowed and planted and got by somehow. The children did what they could to help, and when I am thru with our "claim" it will go to them.

Mrs. N. E. Cannon
Box 535
Butler, Okla.

Jonas and Papa's Pig

My parents lived in the Creek Nation, part of the Indian Territory, before Oklahoma became a state. The Indians were friendly, altho they resented the fact that white people had spoiled their hunting grounds. They felt entitled to food they might steal from the whites.

The first years were hard for settlers and Indians alike. My father had bought a small pig in the early spring and had nourished it carefully all summer with table scraps and waste milk. He planned to fatten it in the fall on the acorns from the jack-oak woods. On that little pig the meat and lard supply for the family depended.

The first days of October found the pig in good condition, weighing about 250 pounds. Each day he would forage in the woods and at night he would return.

"As soon as the weather turns cold, we'll butcher that hog," Father promised. How we all looked forward to the event! Fresh meat was very scarce, for not every family had a hog. Neighbors miles away knew they would have a treat, a "mess of meat," whenever another homesteader butchered.

Then one evening the hog failed to come home! Mother was

worried at once, but Father said, "Now, Mother, the acorns are falling faster and he isn't hungry enough to hurry home anymore. Besides, he's earmarked with my mark—a crop, a split and an underbit in the right ear!"

"I know," answered Mother, "but that earmark won't help you a bit after the ear is eaten!" So Father started out to hunt for the hog.

First, he went to town, which was just a general store, post office and loafing place combined. Several Indians were there. Among them was Jonas, a good friend of my father. They visited awhile, and then Father remarked that he would like to buy a fat hog. Jonas said he had one to sell. He and Father mounted their ponies and rode out to Jonas' place.

They rode up to the pigpen, an unusual thing in those days because all hogs foraged in the woods.

There lay our precious winter's meat supply, grunting contentedly. Father pointed out the earmark and said, "Jonas, that is my hog!"

"Huh," grunted Jonas, not at all perturbed, "if pig yours, better take him home." With no sign of chagrin, he dismounted and opened the gap in the pen and our winter's wealth trotted home.

Mrs. W. W. Dunaway
914 W. Padon
Blackwell, Okla.

Watched Indian Dances

We went to the Cheyenne and Arapaho country when western Oklahoma was new. It was quite a trip from Nebraska. It was February and very cold. There was one terrible blizzard in Kansas. We made it at last and landed on our claim.

Father made us a house out of split oak timber stood on end and with the cracks daubed with mud. There were two small windows and a door. It was 14 by 16 with a small garret overhead where we children slept.

After we finished the house, Father left Mother and us children and went 100 miles away to find work to buy us something to eat. Wages were from 50 cents to a dollar a day for a man and work was scarce at that. For fruit we gathered wild plums and currants and sweetened them with sorghum.

There were lots of deer, and big gray wolves were plentiful. Indians would go by in droves a mile or two long going to Indian dances. They would ride in lumber wagons and on horseback.

The Indian who could dance the longest was the bravest. I saw a boy about 12 years old dance until he fell like he was dead. The Indians gathered around him and laughed and praised him.

Those were happy days and everybody enjoyed the same fare.

Mrs. Ethel Tichenor
Neosho Falls, Kan.

Thought He Was a Lawyer

My father made the Run into the Cherokee Strip when that part of Oklahoma was opened for settlement. Mother and their two-year-old baby girl followed with a wagon load of supplies. Mother and Father still own the land they homesteaded.

Mother, who is now 86, and Father, who is 88, say life on the homestead was good, but sometimes very lonely. They were so eager for company that even when they had nothing to eat but beans, cornbread and sirup, they took some family home from Sunday School with them so they could have someone to talk to. Dad was superintendent of the Sunday School and Mother was the organist.

Mother was a good practical nurse. Many times she would tuck us in at night, but the next morning when we woke we would have no mother. She had gone in the night to care for a sick person on some other lonely homestead.

Dad had a book on homesteading laws, and men would come to him for miles around to get information from his law book. Many people thought he was a lawyer just because he owned this book.

Mother and Dad celebrated their 65th wedding anniversary last fall. They're still pioneers at heart—working together to seek out new ways to do kind things for others!

Lillian Compton
Hobart, Okla.

The Hard Years

I was married in 1891, and my husband made the race and got a claim when the Cherokee Strip opened in 1893. In March, 1894, we moved and lived in a dugout the first year. The walls and floor were dirt. There was a lot of timber just two miles from us, so when a sawmill came in we got green lumber and boarded up the sides of the dugout and put in a floor.

I bought old newspapers at the printing press at Kingman and papered the walls, and got gunny sacks for a carpet in half of the room. When I finished I thought my room was beautiful!

The first year was the hardest. We had no grain for the horses, so my husband would break sod with a walking plow in the mornings and then turn the horses out to grass and plant corn with a hand planter in the afternoons.

The corn looked fine when my husband went back to Kansas to help with the harvest. We had a little girl, two, and she and I stayed alone in the dugout for five weeks. Our corn began to burn up, so I cut it for fodder with a corn knife. I had 25 big shocks. We had a neighbor who went to Kansas and shucked corn for a cent and a half a bushel. He brought enough corn home with him to fatten a hog. That was the best meat I ever tasted!

In the meantime we had another little girl. When she was eight months old and the other little girl was four we lost them both with diphtheria. One died on the 17th of January and the other on the 18th in 1896. We put them both in the same grave. They were the only children we had then.

Mrs. Z. Smith
Wakita, Okla.

E

F

G

H